Building an Asterisk GUI in C#

T.R. Lewis

ISBN-13:

978-1530169344

ISBN-10:

1530169348

DEDICATION

This book is the results of over 18 years VoIP experience and many hours of hard work and determination. For that I dedicate this book to those that pushed me to go the extra mile and kept me from ever giving up.

Contents

Legal

All Software and Operating systems are open source are readily available online along with the user agreements.

I make no assumptions that any information contained within this book is my original work or unique in any manner only that the configuration and combinations are unique.

This book is a reference to information available however the methods used within this book are the methods I have developed over many years of trial and error.

This book cannot be duplicated or copied in any manner without written approval of the author.

Asterisk is a trademark as well as Digium and I make no assumptions that either company endorses this publication.

Redhat, CentOs and Fedora EUA are available at their respected sites I can't do all of the work for you.

Why I wrote this book

I have been working within the VoIP industry for over 18 years with the bulk of it being wrapped around Asterisk and other early open source projects. I have an extensive background in software development and all major programming languages on all major operating systems including early development of Communication Protocols, Telecom and TCP/IP protocols. I've said this so that you may have a little bit of a background on just who I am.

I've chosen to simply write and freelance VoIP support these days and I find myself in the same situations all the time. I usually get a call from someone who thinks that they have installed an Asterisk system and thinks that the quick easy way was the way to go and all should work fine. There are many issues with this line of thought and with the business model that was in place when they installed the system. Now I'm not saying that everyone that calls me are idiots, the bulk of them have made extremely stupid mistakes and I will explain why.

Biggest mistake that is made is they think like an end user and love windows, yes even the ones that run a flavor of Linux on their laptops love windows. How can I say that? Well it's easy they have installed a Linux system with a GUI (Graphical User Interface) to make it look and feel like a windows operating system so that they can easily find their way around. They then brag to their friends that they run Linux and think that makes them special. Fact is they are special but not in a good way.

End Users also like to think that it's ok to run several major services / systems on a single computer and that VoIP will play nice with it. Fact is if it wasn't for VoIP that train of thought might work and has with larger servers coming out every day with more and more memories that is becoming a standard.

End users love the windows feature of click to install, using an EXE or MSI. The make the assumption that all systems are the same. That's because they have usually either grown up in a world where end users are restricted to what they can do (Apple and IPhone) so as to protect the systems running the devices. Apple made a fortune treating end users as idiots and the yuppies enjoy being treated that way. They didn't care that they couldn't access any real system features or that any software written for the devices had to comply to Apple's idiotic standards.

In the following chapters we will address these issues and more as well as explain why this thinking was incorrect and how to do it correctly in the future, we will discuss all relative information from installation to final product of a carrier class system that can be maintained and trusted

Getting Started

In this book we presume that you have built an Asterisk server already and use Realtime as the configuration setup for the Asterisk server. If you have not, then I recommend getting my other book "High Availability Asterisk" from Amazon as well.

We will be going over API's as well as databases to administrate and monitor the Asterisk system and MySql connector with standard queries. However, in this book we will just introduce you to the API's and in the next developer book we will go more into details about them.

If you find yourself wanting consulting or further information, then feel free to email me direct at AsteriskHA@3states.net any time and I will do my best to answer your questions ASAP

Asterisk

Asterisk is a software implementation of a telephone private branch exchange (PBX); it allows attached telephones to make calls to one another, and to connect to other telephone services, such as the public switched telephone network (PSTN) and Voice over Internet Protocol (VoIP) services. Its name comes from the asterisk symbol, *.

Asterisk is released with a dual license model, using the GNU General Public License (GPL) as a free software license and a proprietary software license to permit licensees to distribute proprietary, unpublished system components.

Asterisk was created in 1999 by Mark Spencer of Digium. Originally designed for Linux, Asterisk runs on a variety of operating systems, including NetBSD, OpenBSD, FreeBSD, Mac OS X, and Solaris. Asterisk is small enough to run in an embedded environment such as Customer-premises equipment-hardware running OpenWrt, There are complete self-contained versions that can boot from a storage device such as a flash drive or external disk drive (preferably IDE/PATA, SATA or mSATA; a USB-connected device can be used, but is often not recommended). A live CD or virtual machine can also be used.

The Asterisk software includes many features available in proprietary PBX systems: voice mail, conference calling, interactive voice response (phone menus), and automatic call distribution. Users can create new functionality by writing dial plan scripts in several of Asterisk's own *extensions* languages, by adding custom loadable modules written in C, or by implementing *Asterisk Gateway Interface* (AGI) programs using any programming language capable of communicating via the standard streams system (stdin and stdout) or by network TCP sockets.

Asterisk supports several standard voice over IP protocols, including the

Session Initiation Protocol (SIP), the Media Gateway Control Protocol (MGCP), and H.323. Asterisk supports most SIP telephones, acting both as registrar and back-to-back user agent, and can serve as a gateway between IP phones and the public switched telephone network (PSTN) via T- or E-carrier interfaces or analog FXO cards. The Inter-Asterisk eXchange (IAX) protocol, RFC 5456, native to Asterisk, provides efficient trunking of calls among Asterisk PBXes, in addition to distributing some configuration logic. Many VoIP service providers support it for call completion into the PSTN, often because they themselves have deployed Asterisk or offer it as a hosted application. Some telephones also support the IAX protocol.

By supporting a variety of traditional and VoIP telephony services, Asterisk allows deployers to build telephone systems, or migrate existing systems to new technologies. Some sites are using Asterisk to replace proprietary PBXes, others provide additional features, such as voice mail or voice response menus, or virtual call shops, or to reduce cost by carrying long-distance calls over the Internet (toll bypass).

In addition to VoIP protocols, Asterisk supports traditional circuit-switching protocols such as ISDN and SS7. This requires appropriate hardware interface cards, marketed by third-party vendors. Each protocol requires the installation of software modules.

Asterisk-based telephony solutions offer a rich and flexible feature set. Asterisk offers both classical PBX functionality and advanced features, and interoperates with traditional standards-based telephony systems and Voice over IP systems. Asterisk offers the advanced features that are often associated with large, high end (and high cost) proprietary PBXs. The list below includes a sample of the features available in Asterisk.

MySql

MySQL is one of the most reliable and most recognized database systems out there as well as having most user friendly programming APIs available on the market.

MySql has great connectors that are available for download from their website.

We will create stored routines in the MySql server to keep from having to continually having to write queries.

1) ChangeVMPass()

```
CREATE DEFINER=`root`@`%` PROCEDURE `ChangeVMPAss`(IN
`CustMB` VARCHAR(50), IN `NewPW` VARCHAR(50))
LANGUAGE SQL
NOT DETERMINISTIC
CONTAINS SQL
SQL SECURITY DEFINER
COMMENT ''
BEGIN
UPDATE voicemail_users SET password=NewPW WHERE
mailbox=CustMB;
END
```

2) CreaateNewPhone()

```
CREATE DEFINER=`root`@`%` PROCEDURE
`CreateNewPhone`(IN `IName` VARCHAR(50), IN
`IAccountCode` VARCHAR(50), IN `IDescription`
VARCHAR(50))
LANGUAGE SQL
NOT DETERMINISTIC
CONTAINS SQL
SQL SECURITY DEFINER
```

```
COMMENT ''
BEGIN
INSERT INTO sip_buddies
(name,callerid,defaultuser,accountcode,callbackextens
ion,Description)
VALUES
(IName,CONCAT('"',IDescription,'"
','<',IName,'>'),IName,IAccountCode,IName,IDescriptio
n);
END
```

3) CreateVMBox()

```
CREATE DEFINER=`root`@`%` PROCEDURE `CreateVMBox`(IN
`IAccountCode` VARCHAR(50), IN `IName` VARCHAR(50),
IN `IBox` VARCHAR(50), IN `IPass` VARCHAR(50))
LANGUAGE SQL
NOT DETERMINISTIC
CONTAINS SQL
SQL SECURITY DEFINER
COMMENT ''
BEGIN
UPDATE sip_buddies SET mailbox =
CONCAT(IBox,'@',IAccountCode)
WHERE name=IName;

INSERT INTO voicemail_users
(customer_id,context,mailbox,password,fullname)
VALUES(IName,IAccountCode,IBox,IPass,
(SELECT Description FROM sip_buddies WHERE
name=IName));

INSERT INTO extensions
(context,exten,priority,app,appdata)
VALUES('from-sip-
external1',IName,'1','Dial','SIP/${EXTEN},26');
INSERT INTO extensions
(context,exten,priority,app,appdata)
VALUES('from-sip-
external1',IName,'2','VoiceMail',CONCAT(IBox,'@',IAcc
ountCode));
END
```

4) GetAllCustInfo()

```
CREATE DEFINER=`root`@`%` PROCEDURE
`GetAllCustInfo`(IN `CustomerID` VARCHAR(50))
LANGUAGE SQL
NOT DETERMINISTIC
CONTAINS SQL
SQL SECURITY DEFINER
COMMENT ''
BEGIN
SELECT * FROM sip_buddies WHERE Name=CustomerID;
END
```

5) GetCustomerInfo()

```
CREATE DEFINER=`root`@`%` PROCEDURE
`GetCustomerInfo`()
LANGUAGE SQL
NOT DETERMINISTIC
CONTAINS SQL
SQL SECURITY DEFINER
COMMENT ''
BEGIN
SELECT accountcode FROM sip_buddies GROUP BY
accountcode;
END
```

6) GetCustomerNumberFromAccountCode()

```
CREATE DEFINER=`root`@`%` PROCEDURE
`GetCustomerNumberFromAccountCode`(IN `CustomerID`
VARCHAR(50))
LANGUAGE SQL
NOT DETERMINISTIC
CONTAINS SQL
SQL SECURITY DEFINER
COMMENT ''
BEGIN
```

```
SELECT name FROM sip_buddies WHERE
accountcode=CustomerID;
END
```

7) GetVMinfo()

```
CREATE DEFINER=`root`@`%` PROCEDURE `GetVMInfo`(IN
`CustomerID` VARCHAR(50))
LANGUAGE SQL
NOT DETERMINISTIC
CONTAINS SQL
SQL SECURITY DEFINER
COMMENT ''
BEGIN
SELECT mailbox, password FROM voicemail_users
WHERE mailbox=(SELECT
LEFT(mailbox,LOCATE('@',mailbox) - 1) AS mailbox
FROM sip_buddies Where name=CustomerID);
END
```

8) NewCustomer()

```
CREATE DEFINER=`root`@`%` PROCEDURE `NewCustomer`(IN
`IName` VARCHAR(50), IN `IMailbox` VARCHAR(50), IN
`IAccountcode` VARCHAR(50), IN `IDescription`
VARCHAR(50))
LANGUAGE SQL
NOT DETERMINISTIC
CONTAINS SQL
SQL SECURITY DEFINER
COMMENT ''
BEGIN
INSERT INTO sip_buddies
(name,callerid,defaultuser,mailbox,accountcode,callba
ckextension,Description)
VALUES
(IName,CONCAT(IName,'@',IDescription),IName,IMailbox,
IAccountcode,IName,IDescription);
END
```

Programming Language Tools

One of the best things about Visual studios is that there are tools available for you. Some tools are free while some cost money, however we only need the free ones here. We will be using C# and doing it my way, which may or may not be the standard methods however they are easy to use and follow.

While Visual Studio will let you click and drag data structures into your program I recommend staying away from being that lazy and actually do some work. In the long run you will not only understand what you have wrote but be able to better debug it if needed.

We may discuss some other options however the best option that I've found is in the source code that I've provided at the end of this book.

And as always feel free to email me for additional help (PUT EMAIL HERE)

Ajax

Ajax is a set of web development techniques utilizing many web technologies used on the client-side to create asynchronous Web applications. With Ajax, web applications can send data to and retrieve from a server asynchronously (in the background) without interfering with the display and behavior of the existing page. By decoupling the data interchange layer from the presentation layer, Ajax allows for web pages, and by extension web applications, to change content dynamically without the need to reload the entire page. Data can be retrieved using the XMLHttpRequest object. Despite the name, the use of XML is not required (JSON is often used in the AJAJ variant), and the requests do not need to be asynchronous.

Ajax is not a technology, but a group of technologies. HTML and CSS can be used in combination to mark up and style information. The DOM is accessed with JavaScript to dynamically display – and allow the user to interact with – the information presented. JavaScript and the XMLHttpRequest object provide a method for exchanging data asynchronously between browser and server to avoid full page reloads.

We use Ajax and update panels from the toolkit available free.

Update Panel is our friend

Ajax update panel is one of the tools that is required for real-time data and we love it. There are other methods that some use with lesser languages such as PHP and most require a page refresh that gets ugly especially under heavy use.

I've included some info as stated from the official AJAX site so you can have an ideal of what we are talking about and why we would use it. There is much more info available for the AJAX platform and the UpdatePanel control and if you get time and want to become a better programmer I suggest that you take the time to read about it.

We will use the update panel in our example as well as in most programs that we create. It makes things look and react much more professional.

UpdatePanel Control Overview

Introduction

ASP.NET UpdatePanel controls enable you to build rich, client-centric Web applications. By using UpdatePanel controls, you can refresh selected parts of the page instead of refreshing the whole page with a postback. This is referred to as performing a partial-page update. A Web page that contains a ScriptManager control and one or more UpdatePanel controls can automatically participate in partial-page updates, without custom client script.

This topic contains information about the following:

Scenarios

Background

There is other info available however we are focused on giving you what you need for our application and not looking to confuse you.

Scenarios

The UpdatePanel control is a server control that helps you develop Web pages with complex client behavior that makes a Web page appear more interactive to the end user. Coordinating between server and client to update only specified parts of a Web page usually requires in-depth knowledge of ECMAScript (JavaScript). However, by using the UpdatePanel control, you can enable a Web page to participate in partial-page updates without writing any client script. If you want, you can add custom client script to enhance the client user experience.

When you use an UpdatePanel control, the page behavior is browser independent and can potentially reduce the amount of data that is transferred between client and server.

Background

UpdatePanel controls work by specifying regions of a page that can be updated without refreshing the whole page. This process is coordinated by the ScriptManager server control and the client PageRequestManager class. When partial-page updates are enabled, controls can asynchronously post to the server. An asynchronous postback behaves like a regular postback in that the resulting server page executes the complete page and control life cycle. However, with an asynchronous postback, page updates are limited to regions of the page that are enclosed in UpdatePanel controls and that are marked to be updated. The server sends HTML markup for only the affected elements to the browser. In the browser, the client PageRequestManager class performs Document Object Model (DOM) manipulation to replace existing HTML with updated markup. The following illustration shows a page that is loaded for the first time, and a subsequent asynchronous postback that refreshes the content of an UpdatePanel control.

API's

API's are our friend and as it turns out Asterisk has a free API that we will be using a lot of in the advanced development book. Here though I've included only as inspiration for you to be creative.

The Asterisk Manager Interface (AMI) allows a client program to connect to an Asterisk instance and issue commands or read events over a TCP/IP stream. Integrators will find this particularly useful when trying to track the state of a telephony client inside Asterisk, and directing that client based on custom (and possibly dynamic) rules.

A simple "key: value" line-based protocol is utilized for communication between the connecting client and the Asterisk PBX. Lines are terminated using CR/LF. For our purposes, we will use the term "packet" to describe a set of "key: value" lines that are terminated by an extra CR/LF.

What that basically means is that you can create a simple telnet connection and capture the results for use in your program. This is easy stuff but real powerful results, check out our advanced programming for Asterisk book for samples of use.

Asterisk Manager API

The Asterisk Manager Interface (AMI) allows a client program to connect to an Asterisk instance and issue commands or read events over a TCP/IP stream (Basic Telnet session). This is particularly useful when trying to track the state of a telephony client inside Asterisk, and directing that client based on custom (and possibly dynamic) rules as well as setting up possible thresholds for alarms.

A simple "key: value" line-based protocol is utilized for communication between the connecting client and the Asterisk PBX. Lines are terminated using CR/LF. For our sake, we will use the term "packet" to describe a set of "key: value" lines that are terminated by an extra CR/LF.

Protocol Behavior

The protocol has the following characteristics:

Protocol Behavior

The protocol has the following characteristics:

- Before issuing commands to Asterisk, you must establish a manager session (see below).
- Packets may be transmitted in either direction at any time after authentication.
- The first line of a packet will have a key of "Action" when sent from the client to Asterisk, but "Event" or "Response" when sent from Asterisk to the client.
- The order of lines within a packet is insignificant, so you may use your favorite programming language's native unordered dictionary type to efficiently store a single packet.
- CR/LF is used to delimit each line and a blank line (two CR/LF in a row) indicates the end of the command which Asterisk is now expected to process.

Packet Types

The type of a packet is determined by the existence of one of the following keys:

- **Action**: A packet sent by the connected client to Asterisk, requesting a particular Action be performed. There are a finite (but extendable) set of actions available to the client, determined by the modules presently loaded in the Asterisk engine. Only one action may be outstanding at a time. The Action packet contains the name of the operation to be performed as well as all required parameters.
- **Response**: the response sent by Asterisk to the last action sent by the client.
- **Event**: data pertaining to an event generated from within the Asterisk core or an extension module.

Generally the client sends Action packets to the Asterisk server, the Asterisk server performs the requested operation and returns the result (often only success or failure) in a Response packet. As there is no guarantee regarding the order of Response packets the client usually includes an ActionID parameter in every Action packet that is sent back by Asterisk in the corresponding Response packet. That way the client can easily match Action and Response packets while sending Actions at any desired rate without having to wait for outstanding Response packets before sending the next action.

Event packets are used in two different contexts: On the one hand they inform clients about state changes in Asterisk (like new channels being created and hung up or agents being logged in and out) on the other hand they are used to transport the response payload for actions that return a list of data (event generating actions). When a client sends an event generating action Asterisk sends a Response packed indicating success and containing a "Response: Follows" line. Then it sends zero or more events that contain the actual payload and finally an action complete event indicating that all data has been sent. The events sent in response to an event generating action and the action complete event contain the ActionID of the Action packet that triggered them, so you can easily match them the same way as Response packets. An example of an event generating action is the Status action that triggers Status events for each active channel.

When all Status events have been sent a terminating a StatusComplete event is sent.

Opening a Manager Session and Authenticating as a User

In order to access the Asterisk Manager functionality a user needs to establish a session by opening a TCP/IP connection to the listening port (usually 5038) of the Asterisk instance and logging into the manager using the 'Login' action. This requires a previously established user account on the Asterisk server. User accounts are configured in /etc/asterisk/manager.conf. A user account consists of a set of permitted IP hosts, an authentication secret (password), and a list of granted permissions.

There is a finite set of permissions, each may be granted with either "read", "write", or "read/write" granularity. If a client is granted the ability to read a given class, Asterisk will send it events of that class. If a client is granted the ability to write a given class, it may send actions of that class.

To login and authenticate to the manager, you must send a "Login" action, with your user name and secret (password) as parameters. Here is an example:

```
Action: login
Username: admin
Secret: thisisme
```

If you do not need to subscribe to events being generated by Asterisk, you may also include the "Events: off" parameter, which will prevent event packets being sent to your connection. This is the equivalent of calling the "Events" action. Example:

```
Action: login
Username: admin
Secret: thisisme
Events: off
```

Action Packets

When sending Asterisk an action, extra keys may be provided to further direct execution, for example, you may wish to specify a number to call, a channel to disconnect. Additionally, if your action causes Asterisk to execute an entry in the dialplan, you may wish to pass variables to the dialplan. This is done exactly the same way you would send keys.

To send Asterisk an action, follow this simple format:

Action: <action type><CRLF>
<Key 1>: <Value 1><CRLF>
<Key 2>: <Value 2><CRLF>
...
Variable: <Variable 1>=<Value 1><CRLF>
Variable: <Variable 2>=<Value 2><CRLF>
...
<CRLF>

Manager Actions

Output from the CLI command manager show commands:

- AbsoluteTimeout: Set Absolute Timeout (privilege: call,all)
- ChangeMonitor: Change monitoring filename of a channel (privilege: call,all)
- Command: Execute Command (privilege: command,all)
- Events: Control Event Flow
- ExtensionState: Check Extension Status (privilege: call,all)
- GetVar: Gets a Channel Variable (privilege: call,all)
- Hangup: Hangup Channel __(privilege: call,all)
- IAXpeers: List IAX Peers (privilege: system,all)
- ListCommands: List available manager commands
- Logoff: Logoff Manager
- MailboxCount: Check Mailbox Message Count (privilege: call,all)

- MailboxStatus: Check Mailbox (privilege: call,all)
- Monitor: Monitor a channel (privilege: call,all)
- Originate: Originate Call (privilege: call,all) NOTE: starting from 1.6: originate,all
- ParkedCalls: List parked calls
- Ping: Ping
- QueueAdd: Queues (privilege: agent,all)
- QueueRemove: Queues (privilege: agent,all)
- Queues: Queues
- QueueStatus: Queue Status
- Redirect: Redirect (privilege: call,all)
- SetCDRUserField: Set the CDR UserField (privilege: call,all)
- SetVar: Set Channel Variable (privilege: call,all)
- SIPpeers: List SIP Peers (chan_sip2 only. Not available in chan_sip as of 9/20/2004) (privilege: system,all)
- Status: Status (privilege: call,all)
- StopMonitor: Stop monitoring a channel (privilege: call,all)
- ZapDialOffhook: Dial over Zap channel while offhook
- ZapDNDoff: Toggle Zap channel Do Not Disturb status OFF
- ZapDNDon: Toggle Zap channel Do Not Disturb status ON
- ZapHangup: Hangup Zap Channel
- ZapTransfer: Transfer Zap Channel
- ZapShowChannels: Show Zap Channels

Asterisk 1.2.1:

- AgentCallbackLogin: Sets an agent as logged in by callback (Privilege: agent,all)
- AgentLogoff: Sets an agent as no longer logged in (Privilege: agent,all)
- Agents: Lists agents and their status (Privilege: agent,all)
- DBGet: Get DB Entry (Privilege: system,all)
- DBPut: Put DB Entry (Privilege: system,all)
- QueuePause: Makes a queue member temporarily unavailable (Privilege: agent,all)

- SIPshowPeer: Show SIP peer (text format) (Privilege: system,all)

New in Asterisk 1.4.0

- GetConfig: Display a configuration file, used mainly by AJAM/Asterisk-gui. (Privilege: config,all)
- PlayDTMF: Play DTMF signal on a specific channel. (Privilege: call,all)
- UpdateConfig: Updates a configuration file, used mainly by AJAM/Asterisk-gui. (Privilege: config,all)

Available in Asterisk 1.6.0

- AbsoluteTimeout: Set Absolute Timeout (Priv: system,call,all)
- AgentLogoff: Sets an agent as no longer logged in (Priv: agent,all)
- Agents: Lists agents and their status (Priv: agent,all)
- AGI: Add an AGI command to execute by Async AGI (Priv: call,all)
- Bridge: Bridge two channels already in the PBX (Priv: call,all)
- Challenge: Generate Challenge for MD5 Auth (Priv: <none>)
- ChangeMonitor: Change monitoring filename of a channel (Priv: call,all)
- Command: Execute Asterisk CLI Command (Priv: command,all)
- CoreSettings: Show PBX core settings (version etc) (Priv: system,reporting,all)
- CoreShowChannels: List currently active channels (Priv: system,reporting,all)
- CoreStatus: Show PBX core status variables (Priv: system,reporting,all)
- CreateConfig: Creates an empty file in the configuration directory (Priv: config,all)
- DAHDIDialOffhook: Dial over DAHDI channel while offhook (Priv: <none>)

- DAHDIDNDoff: Toggle DAHDI channel Do Not Disturb status OFF (Priv: <none>)
- DAHDIDNDon: Toggle DAHDI channel Do Not Disturb status ON (Priv: <none>)
- DAHDIHangup: Hangup DAHDI Channel (Priv: <none>)
- DAHDIRestart: Fully Restart DAHDI channels (terminates calls) (Priv: <none>)
- DAHDIShowChannels: Show status dahdi channels (Priv: <none>)
- DAHDITransfer: Transfer DAHDI Channel (Priv: <none>)
- DBDel: Delete DB Entry (Priv: system,all)
- DBDelTree: Delete DB Tree (Priv: system,all)
- DBGet: Get DB Entry (Priv: system,reporting,all)
- DBPut: Put DB Entry (Priv: system,all)
- Events: Control Event Flow (Priv: <none>)
- ExtensionState: Check Extension Status (Priv: call,reporting,all)
- GetConfigJSON: Retrieve configuration (JSON format) (Priv: system,config,all)
- GetConfig: Retrieve configuration (Priv: system,config,all)
- Getvar: Gets a Channel Variable (Priv: call,reporting,all)
- Hangup: Hangup Channel (Priv: system,call,all)
- IAXnetstats: Show IAX Netstats (Priv: system,reporting,all)
- IAXpeerlist: List IAX Peers (Priv: system,reporting,all)
- IAXpeers: List IAX Peers (Priv: system,reporting,all)
- ListCategories: List categories in configuration file (Priv: config,all)
- ListCommands: List available manager commands (Priv: <none>)
- Login: Login Manager (Priv: <none>)
- Logoff: Logoff Manager (Priv: <none>)
- MailboxCount: Check Mailbox Message Count (Priv: call,reporting,all)
- MailboxStatus: Check Mailbox (Priv: call,reporting,all)
- MeetmeMute: Mute a Meetme user (Priv: call,all)
- MeetmeUnmute: Unmute a Meetme user (Priv: call,all)
- ModuleCheck: Check if module is loaded (Priv: system,all)
- ModuleLoad: Module management (Priv: system,all)

- Monitor: Monitor a channel (Priv: call,all)
- Originate: Originate Call (Priv: originate,all)
- ParkedCalls: List parked calls (Priv: <none>)
- Park: Park a channel (Priv: call,all)
- PauseMonitor: Pause monitoring of a channel (Priv: call,all)
- Ping: Keepalive command (Priv: <none>)
- PlayDTMF: Play DTMF signal on a specific channel. (Priv: call,all)
- QueueAdd: Add interface to queue. (Priv: agent,all)
- QueueLog: Adds custom entry in queue_log (Priv: agent,all)
- QueuePause: Makes a queue member temporarily unavailable (Priv: agent,all)
- QueuePenalty: Set the penalty for a queue member (Priv: agent,all)
- QueueRemove: Remove interface from queue. (Priv: agent,all)
- QueueRule: Queue Rules (Priv: <none>)
- Queues: Queues (Priv: <none>)
- QueueStatus: Queue Status (Priv: <none>)
- QueueSummary: Queue Status (Priv: <none>)
- Redirect: Redirect (transfer) a call (Priv: call,all)
- Reload: Send a reload event (Priv: system,config,all)
- SendText: Send text message to channel (Priv: call,all)
- Setvar: Set Channel Variable (Priv: call,all)
- ShowDialPlan: List dialplan (Priv: config,reporting,all)
- SIPpeers: List SIP peers (text format) (Priv: system,reporting,all)
- SIPshowpeer: Show SIP peer (text format) (Priv: system,reporting,all)
- SIPshowregistry: Show SIP registrations (text format) (Priv: system,reporting,all)
- Status: Lists channel status (Priv: system,call,reporting,all)
- StopMonitor: Stop monitoring a channel (Priv: call,all)
- UnpauseMonitor: Unpause monitoring of a channel (Priv: call,all)
- UpdateConfig: Update basic configuration (Priv: config,all)
- UserEvent: Send an arbitrary event (Priv: user,all)
- VoicemailUsersList: List All Voicemail User Information (Priv: call,reporting,all)

- WaitEvent: Wait for an event to occur (Priv: <none>)

Authorization for various classes

Read authorization permits you to receive asynchronous events, in general.
Write authorization permits you to send commands and get back responses. The following classes exist:

- system - General information about the system and ability to run system management commands, such as Shutdown, Restart, and Reload.
- call - Information about channels and ability to set information in a running channel.
- log - Logging information. Read-only.
- verbose - Verbose information. Read-only.
- agent - Information about queues and agents and ability to add queue members to a queue.
- user - Permission to send and receive UserEvent.
- config - Ability to read and write configuration files.
- command - Permission to run CLI commands. Write-only.
- dtmf - Receive DTMF events. Read-only.
- reporting - Ability to get information about the system.
- cdr - Output of cdr_manager, if loaded. Read-only.
- dialplan - Receive NewExten and VarSet events. Read-only.
- originate - Permission to originate new calls. Write-only.

Asterisk Manager: Events

Agent Status Events

'Agentcallbacklogin' Event

Description:
[derived from chan_agent.c]

Data Sample:
Event: Agentcallbacklogin
Agent: <agent>
Loginchan: <loginchan>
Uniqueid: <uniqueid>

'Agentcallbacklogoff' Event

Description:
[derived from chan_agent.c]

Data Sample:
Event: Agentcallbacklogoff
Agent: <agent>
Loginchan: <loginchan>
Logintime: <logintime>
Reason: Autologoff
Uniqueid: <uniqueid>

Event: Agentcallbacklogoff
Agent: <agent>
Loginchan: <loginchan>
Logintime: <logintime>
Uniqueid: <uniqueid>

'AgentCalled' Event

Description:
[derived from app_queue.c]

Data Sample:
Event: AgentCalled
AgentCalled: <channel>
ChannelCalling: <channel>
CallerID: <callerid>
Context: <context>
Extension: <extension>
Priority: <priority>

'AgentComplete' Event

Description:
[derived from app_queue.c]

Data Sample:
Event: AgentComplete
Queue: <queue>
Uniqueid: <uniqueid>
Channel: <channel>
Member: <member>
MemberName: <membername>
HoldTime: <holdtime>
TalkTime: <talktime>
Reason: <reason>

'AgentConnect' Event

Description:
[derived from app_queue.c]

Data Sample:
Event: AgentConnect
Queue: <queue>
Uniqueid: <uniqueid>
Channel: <channel>
Member: <member>
MemberName: <membername>
Holdtime: <holdtime>
BridgedChannel: <bridgedchannel>

'AgentDump' Event
Description:
[derived from app_queue.c]

Data Sample:
Event: AgentDump
Queue: <queue>
Uniqueid: <uniqueid>
Channel: <channel>
Member: <member>
MemberName: <membername>

'Agentlogin' Event
Description:
[derived from chan_agent.c]

Data Sample:
Event: Agentlogin
Agent: <agent>
Channel: <channel>
Uniqueid: <uniqueid>

'Agentlogoff' Event

Description:

[derived from chan_agent.c]

Data Sample:

Event: Agentlogoff

Agent: <agent>

Logintime: <logintime>

Uniqueid: <uniqueid>

'QueueMemberAdded' Event

Description:

1. Sent on Action QueueAdd

[derived from app_queue.c]

Data Sample:

Queue: testing

Location: Agent/AgentId

Membership: dynamic

Penalty: 0

CallsTaken: 0

LastCall: 0

Status: 4

Paused: 1

'QueueMemberPaused' Event

Description:

1. Sent on Action: QueuePause

[derived from app_queue.c]

Data Sample:

Event: QueueMemberPaused

Location: <location>

MemberName: <membername>
Paused: <paused>

'QueueMemberStatus' Event

Description:
[derived from app_queue.c]

Possible values are:
/*! Device is valid but channel didn't know state */

1. define AST_DEVICE_UNKNOWN 0

/*! Device is not used */

1. define AST_DEVICE_NOT_INUSE 1

/*! Device is in use */

1. define AST_DEVICE_INUSE 2

/*! Device is busy */

1. define AST_DEVICE_BUSY 3

/*! Device is invalid */

1. define AST_DEVICE_INVALID 4

/*! Device is unavailable */

1. define AST_DEVICE_UNAVAILABLE 5

/*! Device is ringing */

1. define AST_DEVICE_RINGING 6

/*! Device is ringing *and* in use */

1. define AST_DEVICE_RINGINUSE 7

/*! Device is on hold */

1. define AST_DEVICE_ONHOLD 8

fernando.berretta@voipexperts.com.ar

Data Sample:

Event: QueueMemberStatus
Queue: <queue>
Location: <location>
MemberName: <membername>
Membership: <membership>
Penalty: <penalty>
CallsTaken: <callstaken>
LastCall: <lastcall>
Status: <status>
Paused: <paused>

Command Status Events

Call Status Events

'Cdr' Event

Description:
[derived from cdr_manager.c]

Must be enabled in cdr_manager.conf

[general]
enabled = yes

Data Sample:
Event: Cdr
AccountCode:
Source:
Destination:
DestinationContext:
CallerID:
Channel:

DestinationChannel:
LastApplication:
LastData:
StartTime:
AnswerTime:
EndTime:
Duration:
BillableSeconds:
Disposition:
AMAFlags:
UniqueID:
UserField:

'Dial' Event

Description:
[derived from app_dial.c]

Data Sample:
Event: Dial
Privilege: call,all
Source: Local/900@default-2dbf,2
Destination: SIP/900-4c21
CallerID: <unknown>
CallerIDName: default
SrcUniqueID: 1149161705.2
DestUniqueID: 1149161705.4

'ExtensionStatus' Event

Description:
[derived from manager.c]

Data Sample:
Event: ExtensionStatus
Exten: <ext>
Context: <context>

Status: <state>

'Hangup' Event

Description:

[derived from channel.c]

Data Sample:

Event: Hangup

Channel: SIP/101-3f3f

Uniqueid: 1094154427.10

Cause: 0

Cause Codes

- UNALLOCATED = 1
- NO ROUTE TRANSIT NET = 2
- NO_ROUTE_DESTINATION = 3
- CHANNEL_UNACCEPTABLE = 6
- CALL_AWARDED_DELIVERED = 7
- NORMAL_CLEARING = 16
- USER_BUSY = 17
- NO USER RESPONSE = 18
- NO ANSWER = 19
- CALL REJECTED = 21
- NUMBER CHANGED = 22
- DESTINATION OUT OF ORDER = 27
- INVALID NUMBER FORMAT = 28
- FACILITY REJECTED = 29
- RESPONSE TO STATUS ENQUIRY = 30
- NORMAL UNSPECIFIED = 31
- NORMAL CIRCUIT CONGESTION = 34
- NETWORK OUT OF ORDER = 38
- NORMAL TEMPORARY FAILURE = 41
- SWITCH CONGESTION = 42
- ACCESS INFO DISCARDED = 43
- REQUESTED CHAN UNAVAIL = 44
- PRE EMPTED = 45
- FACILITY NOT SUBSCRIBED = 50
- OUTGOING CALL BARRED = 52

- INCOMING CALL BARRED = 54
- BEARERCAPABILITY NOTAUTH = 57
- BEARERCAPABILITY NOTAVAIL = 58
- BEARERCAPABILITY NOTIMPL = 65
- CHAN NOT IMPLEMENTED = 66
- FACILITY NOT IMPLEMENTED = 69
- INVALID CALL REFERENCE = 81
- INCOMPATIBLE DESTINATION = 88
- INVALID MSG UNSPECIFIED = 95
- MANDATORY IE MISSING = 96
- MESSAGE TYPE NONEXIST = 97
- WRONG MESSAGE = 98
- IE NONEXIST = 99
- INVALID IE CONTENTS = 100
- WRONG CALL STATE = 101
- RECOVERY ON TIMER EXPIRE = 102
- MANDATORY IE LENGTH ERROR = 103
- PROTOCOL ERROR = 111
- INTERWORKING = 127
- NOT DEFINED = 0

'MusicOnHold' Event

Description:

1. Occurs when a channel is placed on hold/unhold and music is played to the caller.

Data Sample:

Event: MusicOnHold
Channel: <Channel ID>
State: <Start/Stop>
Uniqueid: <Unique ID>

'Join' Event

Description:

[derived from app_queue.c]

Data Sample:
Event: Join
Channel: <channel>
CallerID: <callerid|unknown>
Queue: <queuename>
Position: <entryposition>
Count: <queuemembercount>

'Leave' Event

Description:
[derived from app_queue.c]

Data Sample:
Event: Leave
Channel: <channel>
Queue: <queuename>
Count: <queuemembercount>

'Link' Event

Description:

1. Fired when two voice channels are linked together and voice data exchange commences.

Notes:

1. Several *Link* events may be seen for a single call. This can occur when Asterisk fails to setup a *native bridge* for the call. As far as I can tell, this is when Asterisk must sit between two telephones and perform CODEC conversion on their behalf.

Data Sample:

Event: Link
Channel1: SIP/101-3f3f
Channel2: Zap/2-1

Uniqueid1: 1094154427.10
Uniqueid2: 1094154427.11

Note: in current version it replaced by Bridge event, example is
Channel2: SIP/1-1.1.1.1-0002ef99
Bridgestate: Link
Event: Bridge
Privilege: call,all
Uniqueid2: 1393028418.530941
Channel1: SIP/peer-local-0002ef98,
Bridgetype: core
Uniqueid1: 1393028346.530940
Timestamp: 1393028428.618726
CallerID1: 323187134981
CallerID2: 323187134981

'MeetmeJoin' Event

Description:
[derived from app_meetme.c]

Data Sample:
Event: MeetmeJoin
Channel: <channel>
Uniqueid: <uniqueid>
Meetme: <meetme>
Usernum: <usernum>

'MeetmeLeave' Event

Description:
[derived from app_meetme.c]

Data Sample:
Event: MeetmeLeave
Channel: <channel>

Uniqueid: <uniqueid>
Meetme: <meetme>
Usernum: <usernum>

'MeetmeStopTalking' Event

Description:
[derived from app_meetme.c]

Notes:

1. This requires the T option on the meetme application

Data Sample:

Event: MeetmeStopTalking
Privilege: call,all
Channel: SIP/200-ABC1
Uniqueid: 1234567890.1
Meetme: 400
Usernum: 2

'MeetmeTalking' Event

Description:
[derived from app_meetme.c]

Notes:

1. This requires the T option on the meetme application

Data Sample:

Event: MeetmeTalking
Privilege: call,all
Channel: SIP/200-ABC1
Uniqueid: 1234567890.1

Meetme: 400
Usernum: 2

'MessageWaiting' Event
Description:
[derived from app_voicemail.c]

Data Sample:
Event: MessageWaiting
Mailbox: <mailbox>@<context>
Waiting: <count>
New: <number>
Old: <number>

Event: MessageWaiting
Mailbox: <context>
Waiting: <count>

'Newcallerid' Event
Description:
[derived from channel.c]

Data Sample:
Event: Newcallerid
Channel: <channel>
Callerid: <callerid>
Uniqueid: <uniqueid>

'Newchannel' Event
Description:
[derived from channel.c]

Data Sample:

Event: Newchannel
Channel: Zap/2-1
State: Rsrvd
Callerid: <unknown>
Uniqueid: 1094154427.11

Event: Newchannel
Channel: SIP/101-3f3f
State: Ring
Callerid: 101
Uniqueid: 1094154427.10

'Newexten' Event

Description:

1. Fired whenever a pbx function (such as execution of dialplan) occurs

Data Sample:
Event: Newexten
Channel: SIP/101-00c7
Context: macro-ext
Extension: s
Priority: 3
Application: Goto
AppData: s-BUSY
Uniqueid: 1094154321.8

Event: Newexten
Channel: SIP/101-3f3f
Context: local_extensions
Extension: 917070
Priority: 1
Application: AGI
AppData: /etc/asterisk/agi/ks_doorman_pickup.py|channel_up
Uniqueid: 1094154427.10

Event: Newexten
Channel: SIP/101-3f3f
Context: local_extensions
Extension: 917070
Priority: 2
Application: Dial
AppData: Zap/G1/17070
Uniqueid: 1094154427.10

'ParkedCall' Event

Description:
[derived from res_features.c]

Data Sample:
Event: ParkedCall
Exten: <parkexten>
Channel: <channel>
From: <from>
Timeout: <timeout>
CallerID: <callerid>

'Rename' Event

Description:
[derived from channel.c: channel 'rename' event]

Data Sample:
Event: Rename
Oldname: <oldname>
Newname: <newname>
Uniqueid: <uniqueid>

'SetCDRUserField' Event

Description:

[derived from app_setcdruserfield.c]

Data Sample:

'Unlink' Event

Description:

1. Fired when a link between two voice channels is discontinued, for example, just before call completion.

Notes:

1. Several *Unlink* events may be seen for a single call. This can occur when Asterisk fails to setup a *native bridge* for the call. As far as I can tell, this is when Asterisk must sit between two telephones and perform CODEC conversion on their behalf.

Data Sample:

Event: Unlink
Channel1: SIP/101-3f3f
Channel2: Zap/2-1
Uniqueid1: 1094154427.10
Uniqueid2: 1094154427.11
Note: in current version it replaced by Bridge event, example is
Channel2: SIP/1-1.1.1.1-0002ef99
Bridgestate: Unlink
Event: Bridge
Privilege: call,all
Uniqueid2: 1393028418.530941
Channel1: SIP/peer-local-0002ef98,
Bridgetype: core
Uniqueid1: 1393028346.530940
Timestamp: 1393028428.618726
CallerID1: 323187134981
CallerID2: 323187134981

'UnParkedCall' Event

Description:
[derived from res_features.c]

Data Sample:

System Status Events

'Alarm' Event:

Description:
[derived from chan_zap.c]

Data Sample:
Event: Alarm
Alarm: <(Red|Yellow|Blue|No|Unknown)
Alarm|Recovering|Loopback|Not Open|None>
Channel: <channel>

'AlarmClear' Event:

Description:
[derived from chan_zap.c]

Data Sample:
Event: AlarmClear
Channel: <channel>

'DNDState' Event:

Description:
[derived from chan_dahdi.c]

Data Sample:
Event: DNDState
Channel: Zap/1
Status: <enabled|disabled>

'LogChannel' Event

Description:
[derived from logger.c]

Data Sample:
Event: LogChannel
Channel: /var/log/asterisk/messages
Enabled: Yes

Event: LogChannel
Channel: /var/log/asterisk/messages
Enabled: No
Reason: 13 - Permission denied

'PeerStatus' Event

Description:

1. Fired when a peer registers/unregisters with Asterisk
[derived from chan_sip.c, chan_iax2.c]

Data Sample:
Event: PeerStatus
Peer: SIP/2005
PeerStatus: Registered

Event: PeerStatus
Peer: SIP/2005
PeerStatus: Unregistered
Cause: Expired

Event: PeerStatus
Peer: IAX2/2007
PeerStatus: <Lagged|Reachable|Unreachable>
Time: 1000

'Registry' Event

Description:

1. Fired when Asterisk registers with a peer
[derived from chan_sip.c, chan_iax2.c]

Notes:
For an entry like:
register =>
username:password:authname@sip.domain:port/local_contact
Domain would reflect the value of sip.domain

Data Sample:
Event: Registry
Channel: SIP
Domain: sip.domain
Status: Registered

'Reload' Event

Description:

1. Fired when the "RELOAD" console command is executed.
[derived from manager.c]

Data Sample:
Event: Reload
Message: Reload Requested

'Shutdown' Event

Description:
[derived from asterisk.c]

Data Sample:
Event: Shutdown

Shutdown: <Uncleanly|Cleanly>
Restart: <True|False>

User Status Events

'UserEvent' Event

Description:
[derived from app_userevent.c]

Data Sample:
Event: <event>
Channel: <channel>
Uniqueid: <uniqueid>

Event: <event>
Channel: <channel>
Uniqueid: <uniqueid>
<body>

Classes and Why

A class is a construct that enables you to create your own custom types by grouping together variables of other types, methods and events. A class is like a blueprint. It defines the data and behavior of a type. If the class is not declared as static, client code can use it by creating objects or instances which are assigned to a variable. The variable remains in memory until all references to it go out of scope. At that time, the CLR marks it as eligible for garbage collection. If the class is declared as static, then only one copy exists in memory and client code can only access it through the class itself, not an instance variable.

We will use classes to keep from writing the same code over and over. This saves time and is what I believe is a real world best practice. They also keep code in neat readable format that can easily be followed.

Introduction to the C# Language and the .NET Framework

Just so you have a little background on C# below is from Microsoft and lets you know why we love C# and use it primarily in development for web applications as well as windows software.

C# is an elegant and type-safe object-oriented language that enables developers to build a variety of secure and robust applications that run on the .NET Framework. You can use C# to create Windows client applications, XML Web services, distributed components, client-server applications, database applications, and much, much more. Visual C# provides an advanced code editor, convenient user interface designers, integrated debugger, and many other tools to make it easier to develop applications based on the C# language and the .NET Framework.

C# syntax is highly expressive, yet it is also simple and easy to learn. The curly-brace syntax of C# will be instantly recognizable to anyone familiar with C, C++ or Java. Developers who know any of these languages are typically able to begin to work productively in C# within a very short time. C# syntax simplifies many of the complexities of C++ and provides powerful features such as nullable value types, enumerations, delegates, lambda expressions and direct memory access, which are not found in Java. C# supports generic methods and types, which provide increased type safety and performance, and iterators, which enable implementers of collection classes to define custom iteration behaviors that are simple to use by client code. Language-Integrated Query (LINQ) expressions make the strongly-typed query a first-class language construct.

As an object-oriented language, C# supports the concepts of encapsulation, inheritance, and polymorphism. All variables and methods, including the Main method, the application's entry point, are encapsulated within class definitions. A class may inherit directly from one parent class, but it may implement any number of interfaces. Methods that override virtual methods in a parent class require the override keyword as a way to avoid accidental redefinition. In C#, a structure is like a lightweight class; it is a stack-allocated type that can implement interfaces but does not support inheritance.

In addition to these basic object-oriented principles, C# makes it easy to develop software components through several innovative language constructs, including the following:

•Encapsulated method signatures called delegates, which enable type-safe event notifications.

•Properties, which serve as accessors for private member variables.

•Attributes, which provide declarative metadata about types at run time.

•Inline XML documentation comments.

•Language-Integrated Query (LINQ) which provides built-in query capabilities across a variety of data sources.

If you have to interact with other Windows software such as COM objects or native Win32 DLLs, you can do this in C# through a process called "Interop." Interop enables C# programs to do almost anything that a native C++ application can do. C# even supports pointers and the concept of "unsafe" code for those cases in which direct memory access is absolutely critical.

The C# build process is simple compared to C and C++ and more flexible than in Java. There are no separate header files, and no requirement that methods and types be declared in a particular order. A C# source file may define any number of classes, structs, interfaces, and events.

.NET Framework Platform Architecture

C# programs run on the .NET Framework, an integral component of Windows that includes a virtual execution system called the common language runtime (CLR) and a unified set of class libraries. The CLR is the commercial implementation by Microsoft of the common language infrastructure (CLI), an international standard that is the basis for creating execution and development environments in which languages and libraries work together seamlessly.

Source code written in C# is compiled into an intermediate language (IL) that conforms to the CLI specification. The IL code and resources, such as bitmaps and strings, are stored on disk in an executable file called an assembly, typically with an extension of .exe or .dll. An assembly contains a manifest that provides information about the assembly's types, version, culture, and security requirements.

When the C# program is executed, the assembly is loaded into the CLR, which might take various actions based on the information in the manifest. Then, if the security requirements are met, the CLR performs just in time (JIT) compilation to convert the IL code to native machine instructions. The CLR also provides other services related to automatic garbage collection, exception handling, and resource management.

Code that is executed by the CLR is sometimes referred to as "managed code," in contrast to "unmanaged code" which is compiled into native machine language that targets a specific system. The following diagram illustrates the compile-time and run-time relationships of C# source code files, the .NET Framework class libraries, assemblies, and the CLR.

From C# source code to machine execution

Language interoperability is a key feature of the .NET Framework. Because the IL code produced by the C# compiler conforms to the Common Type Specification (CTS), IL code generated from C# can interact with code that was generated from the .NET versions of Visual Basic, Visual C++, or any of more than 20 other CTS-compliant languages. A single assembly may contain multiple modules written in different .NET languages, and the types can reference each other just as if they were written in the same language.

In addition to the run time services, the .NET Framework also includes an extensive library of over 4000 classes organized into namespaces that provide a wide variety of useful functionality for everything from file input and output to string manipulation to XML parsing, to Windows Forms controls. The typical C# application uses the .NET Framework class library extensively to handle common "plumbing" chores.

Tips

I am not going into great detail here except to tell you that error logging is good and invaluable to the developer; it gives us the details we need to build better applications.

Some basic rules to live by;

1) Name voids accordingly
2) Name controls accordingly
3) Name classes accordingly
4) Name strings accordingly
5) Always use error handlers
6) Write errors to database whenever possible

If you find yourself wanting consulting or further information, then feel free to email me direct at AsteriskHA@3states.net any time and I will do my best to answer your questions ASAP

Sample Program

Let's start with the basic, simply making a project and adding to it. You must have the stored routines that are in the MySql section of this book in your MySql server. You also must have an Asterisk system as setup in the first book "High Availability Asterisk" found on Amazon as well.

If you find yourself wanting consulting or further information, then feel free to email me direct at AsteriskHA@3states.net any time and I will do my best to answer your questions ASAP

Ok in this example you will;

1) Create a new C# blank web application.
2) Create a class
3) Create functions and Methods
4) Create an interface that permits you to add
    a. Phones
    b. Voicemail
    c. Change Voicemail passwords
5) Create an interface that will let you see current status of the Asterisk system is added at the end of the book under bonus.

Create the Project

Open visual studios and select new project, then select new C# blank web project and name it "AsteriskManager", it is important that the names I use you use as well for this sample, any variations will keep it from working including CAPs and Lowercases.

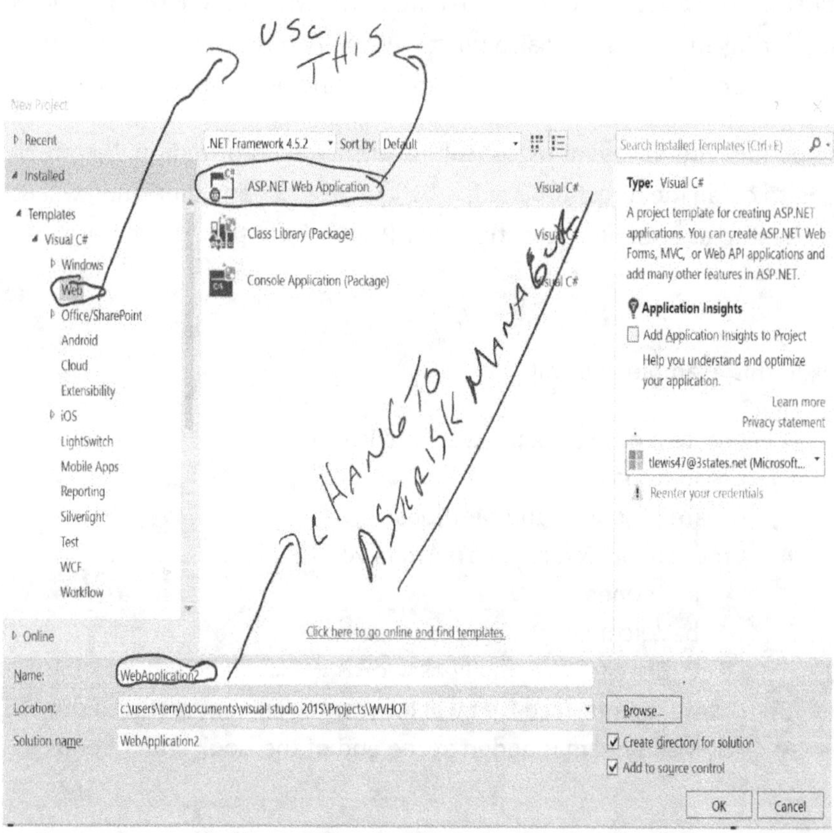

The solution name should automatically change as well.

Then

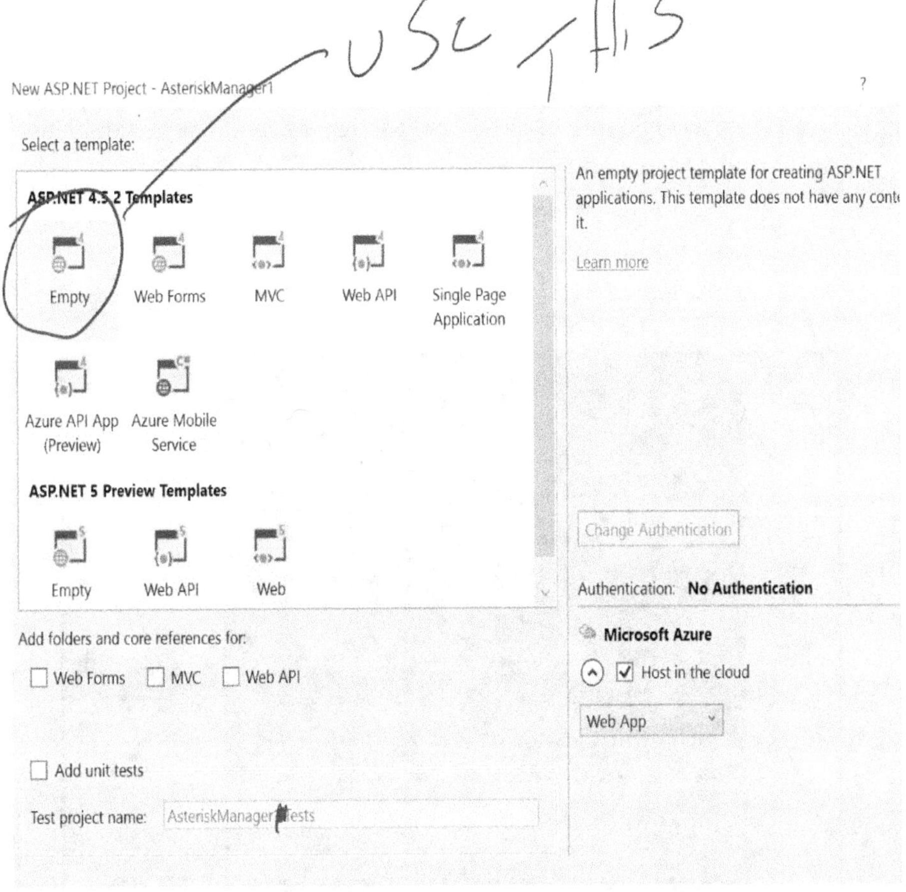

I know the option is there for a "Web Forms" and could spend hours explaining why I don't use it, but I'm not, in just telling you to do it my way.

Then

Make sure you're in the solution explorer and right click on the project name (I know my name has a "1" on it so ignore that, yours should just say "AsteriskManager".

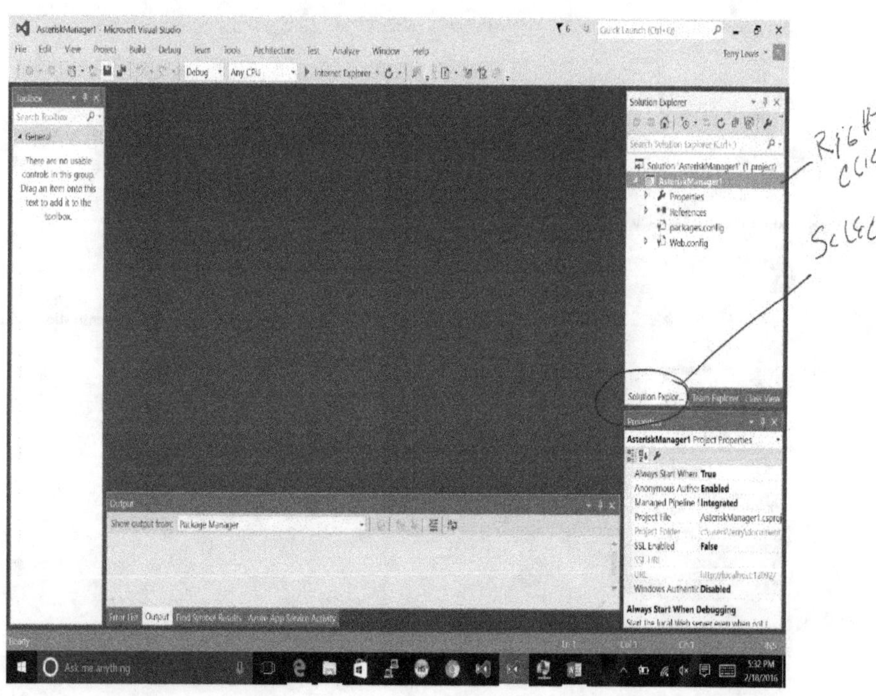

Adding the web form

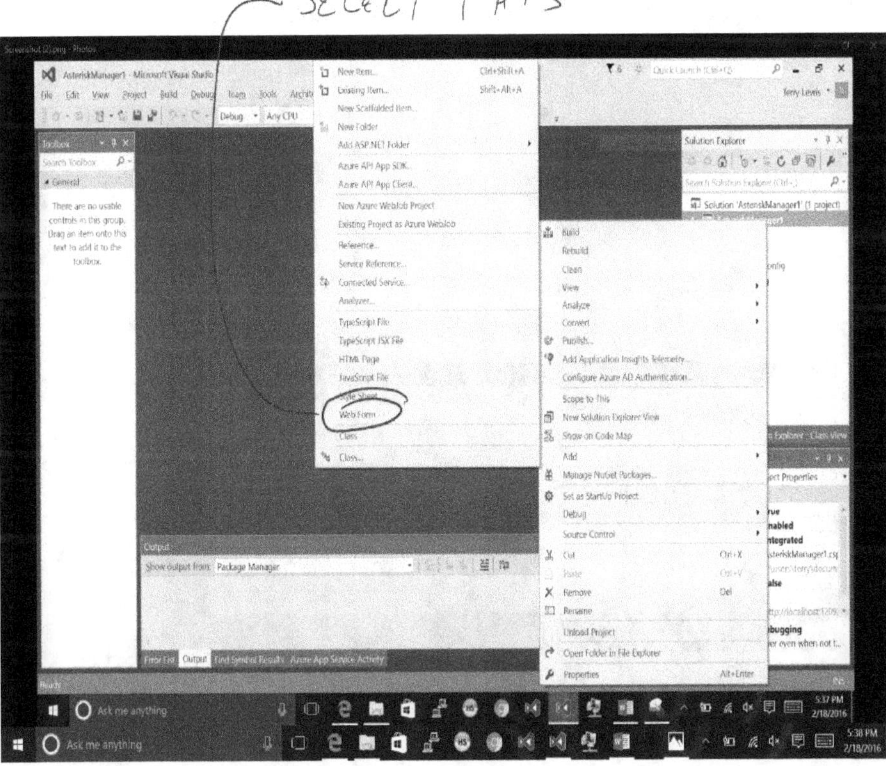

And name it “Default”, this will be the only web form needed for this project sample.

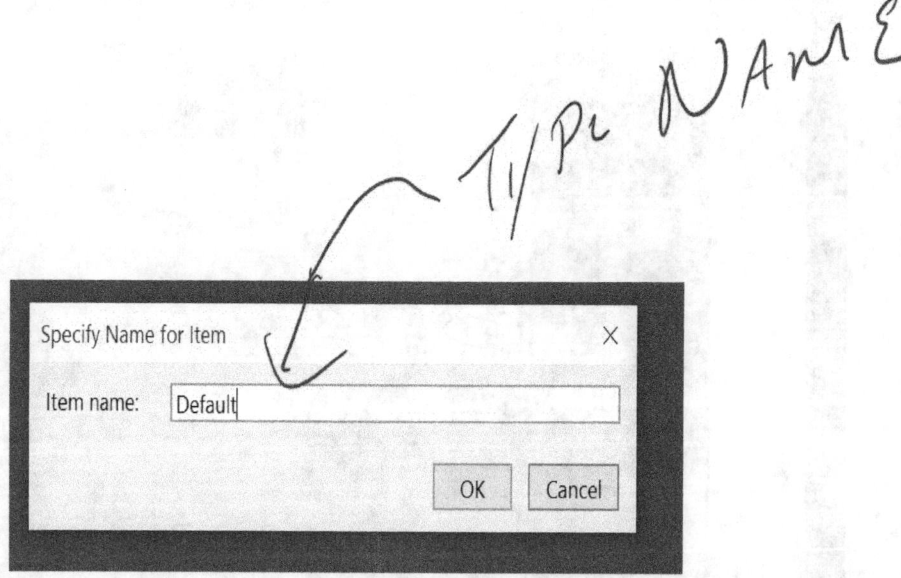

Adding Classes

We will use a clase called “Database” for our sample program.

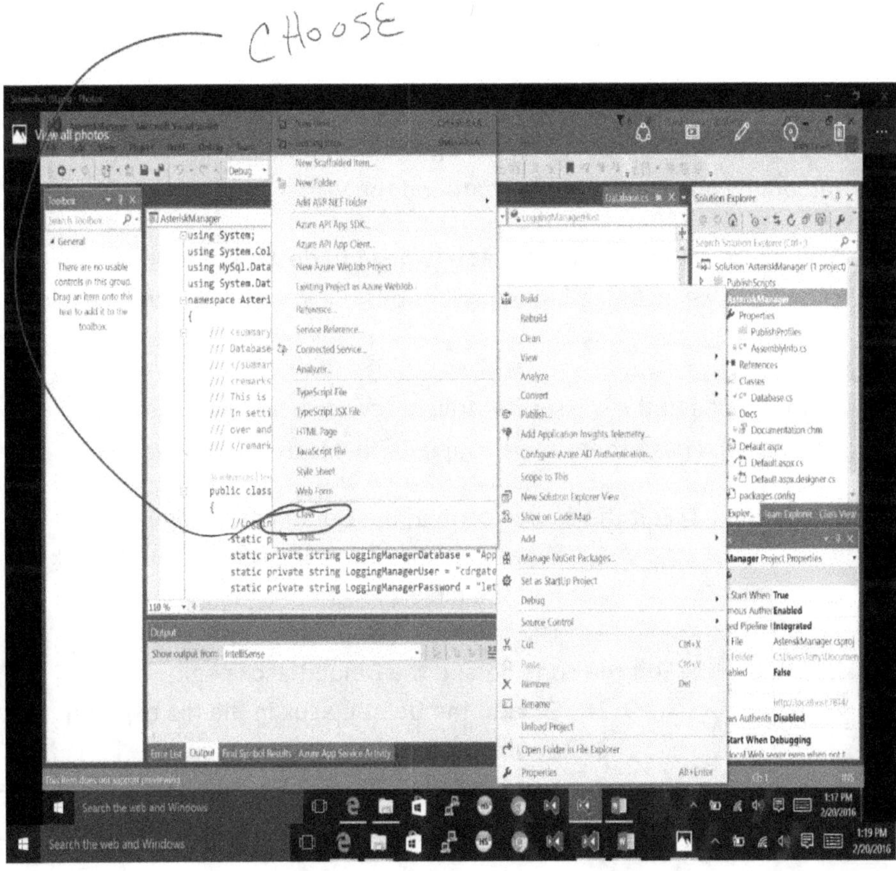

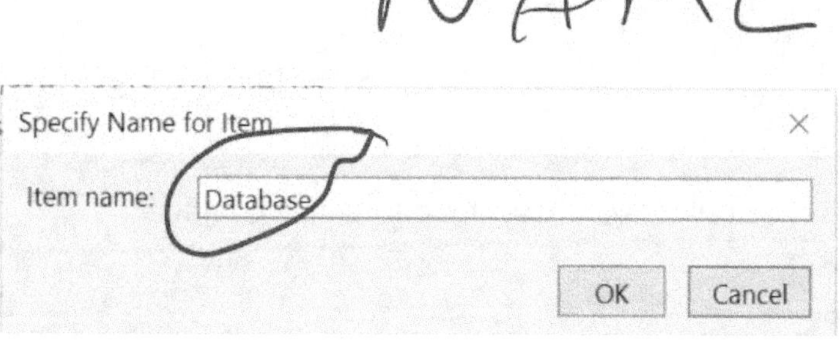

Now we've added our forms and our classes, whats next? Code you idiot, ok well ill let that pass. Let's go to the defauilt.aspx page (assuming you know your way around the Visual Studios RAD that is.

Now there is the issue here that you need to do a lot of typing using the code that I've included in the back of this book make your program look like mine.

With regards to the classes its straight forward, however with the Default.aspx page there are a couple of things that you need to do.

1) Edit the source code of the Default.aspx page
    1. When selecting the page simply double click on it in the file explorer.
    2. Then select the source under the page.
2) Edit the code behind the Default.aspx page
    1. Expand the Default.aspx in the file explorer.
    2. Double-click on the Default.aspx.cs
    3. Edit the code in the window to match what ive included.

Now that seems easy doesn't it? Heres what we have done.

1) Created database routines
    a. They do the work for us.
2) Created a Class in the program

   a. These call the database routines and returns or sets the data as needed.
3) Created the webpage
   a. This is what delivers the data as needed either to the classes or to the end user.

Not rocket science and fairly straight forward and clean. From here you should be able to see what does what as well as think of other things you can manage in the same methods. I only made the program do basic actions, I would recommend adding security to the program before making it live as well as a menu page, doing so will protect you as well as permit you to add future features to the app witout changing the end users access to the program while keeping uniformity to the program.

Keep in mind, end users lie, that's why even though we don't use the logging features here, I have included them. The logging features makes it so you don't have to pay attention to the enduser and focus on the real what, when, where and how things happened with facts not enduser translations of their views.

Project References

DEFAULT CLASS

Partial class Default.

Partial class Default.

◢Inheritance Hierarchy

System. Object
  System.Web.UI. Control
    System.Web.UI. TemplateControl
      System.Web.UI. Page
        AsteriskManager. Default

Namespace: AsteriskManager
Assembly: AsteriskManager (in AsteriskManager.dll) Version: 1.0.0.0 (1.0.0.0)

◢Syntax

C#

VB

C++

F#

Copy

```
public class Default : Page
```

The Default type exposes the following members.

Constructors

	Name	Description
	Default	Initializes a new instance of the Default class

Top

Methods

	Name	Description
	AddContentTemplate	Called during page initialization to create a collection of content (from content controls) that is handed to a master page, if the current page or master page refers to a master page. (Inherited from Page.)

AddedControl	Called after a child control is added to the Controls collection of the Control object. (Inherited from Control.)
AddOnPreRenderCompleteAsync(BeginEventHandler, EndEventHandler)	Registers beginning and ending event handler delegates that do not require state information for an asynchronous page. (Inherited from Page.)
AddOnPreRenderCompleteAsync(BeginEventHandler, EndEventHandler, Object)	Registers beginning and ending event handler delegates for an asynchronous page. (Inherited from Page.)
AddParsedSubObject	Notifies the server control that an element, either XML or HTML, was parsed, and

		adds the element to the server control's ControlCollection object. (Inherited from Control.)
	AddWrappedFileDependencies	Adds a list of dependent files that make up the current page. This method is used internally by the ASP.NET page framework and is not intended to be used directly from your code. (Inherited from Page.)
	ApplyStyleSheetSkin	Applies the style properties defined in the page style sheet to the control. (Inherited from Control.)

	AspCompatBeginProcessRequest	Initiates a request for Active Server Page (ASP) resources. This method is provided for compatibility with legacy ASP applications. (Inherited from Page.)
	AspCompatEndProcessRequest	Terminates a request for Active Server Page (ASP) resources. This method is provided for compatibility with legacy ASP applications. (Inherited from Page.)
	AsyncPageBeginProcessRequest	Begins processing an asynchronous page request. (Inherited from Page.)
	AsyncPageEndProcessRequest	Ends processing an asynchronous page

		request. (Inherited from Page.)
	BuildProfileTree	Gathers information about the server control and delivers it to the Trace property to be displayed when tracing is enabled for the page. (Inherited from Control.)
	ButtonCancel_Click	ButtonCancel_Click.
	ButtonChangeVMPass_Click	ButtonChangeVMPass_Click.
	ButtonCreateNewUser_Click	ButtonCreateNewUser_Click.
	ButtonCreateVM_Click	ButtonCreateVM_Click.
	CheckBoxCurrentCustomer	CheckBoxCurrentCusto

	_CheckedChanged	mer_CheckedChanged.
	ClearCachedClientID	Sets the cached ClientID value to null. (Inherited from Control.)
	ClearChildControlState	Deletes the control-state information for the server control's child controls. (Inherited from Control.)
	ClearChildState	Deletes the view-state and control-state information for all the server control's child controls. (Inherited from Control.)
	ClearChildViewState	Deletes the view-state information for all the server control's child

	Name	Description
		controls. (Inherited from Control.)
	ClearEffectiveClientIDMode	Sets the ClientIDMode property of the current control instance and of any child controls to Inherit. (Inherited from Control.)
	Construct	Performs design-time logic. (Inherited from TemplateControl.)
	CreateChildControls	Called by the ASP.NET page framework to notify server controls that use composition-based implementation to create any child controls they contain in preparation for posting back or

		rendering. (Inherited from Control.)
	CreateControlCollection	Creates a new ControlCollection object to hold the child controls (both literal and server) of the server control. (Inherited from Control.)
	CreateHtmlTextWriter	Creates an HtmlTextWriter object to render the page's content. (Inherited from Page.)
	CreateResourceBasedLiteralControl	Accesses literal strings stored in a resource. The CreateResourceBasedLiteralControl(Int32, Int32, Boolean) method is not

	Name	Description
		intended for use from within your code. (Inherited from TemplateControl.)
	CustomerId_SelectedIndexChanged	CustomerId_SelectedIndexChanged.
	CustomerNumberDL_SelectedIndexChanged	CustomerNumberDL_SelectedIndexChanged.
	CustomerNumberTXT_TextChanged	CustomerNumberTXT_TextChanged.
	DataBind()	Binds a data source to the invoked server control and all its child controls. (Inherited from Control.)
	DataBind(Boolean)	Binds a data source to the invoked server control and all its child controls with an

	option to raise the DataBinding event. (Inherited from Control.)
DataBindChildren	Binds a data source to the server control's child controls. (Inherited from Control.)
DesignerInitialize	Performs any initialization of the instance of the Page class that is required by RAD designers. This method is used only at design time. (Inherited from Page.)
DeterminePostBackMode	Returns a NameValueCollection of data posted back to the page using either a POST or a GET

	command. (Inherited from Page.)
Dispose	Enables a server control to perform final clean up before it is released from memory. (Inherited from Control.)
EnsureChildControls	Determines whether the server control contains child controls. If it does not, it creates child controls. (Inherited from Control.)
EnsureID	Creates an identifier for controls that do not have an identifier assigned. (Inherited from Control.)

	Equals	Determines whether the specified Object is equal to the current Object. (Inherited from Object.)
	Eval(String)	Evaluates a data-binding expression. (Inherited from TemplateControl.)
	Eval(String, String)	Evaluates a data-binding expression using the specified format string to display the result. (Inherited from TemplateControl.)
	ExecuteRegisteredAsyncTasks	Starts the execution of an asynchronous task. (Inherited from Page.)

	fill_Drop_Down	fill_Drop_Down.
	Finalize	Allows an object to try to free resources and perform other cleanup operations before it is reclaimed by garbage collection. (Inherited from Object.)
	FindControl(String)	Searches the page naming container for a server control with the specified identifier. (Inherited from Page.)
	FindControl(String, Int32)	Searches the current naming container for a server control with the specified *id* and an integer, specified in the *pathOffset* parameter, which aids in the search. You should not override

		this version of the FindControl() method. (Inherited from Control.)
	Focus	Sets input focus to a control. (Inherited from Control.)
	FrameworkInitialize	Initializes the control tree during page generation based on the declarative nature of the page. (Inherited from Page.)
	GetDataItem	Gets the data item at the top of the data-binding context stack. (Inherited from Page.)
	GetDesignModeState	Gets design-time data for a control.

		(Inherited from Control.)
	GetGlobalResourceObject(String, String)	Gets an application-level resource object based on the specified ClassKey and ResourceKey properties. (Inherited from TemplateControl.)
	GetGlobalResourceObject(String, String, Type, String)	Gets an application-level resource object based on the specified ClassKey and ResourceKey properties, object type, and property name of the resource. (Inherited from TemplateControl.)
	GetHashCode	Serves as a hash function for a particular type.

		(Inherited from Object.)
	GetLocalResourceObject(String)	Gets a page-level resource object based on the specified ResourceKey property. (Inherited from TemplateControl.)
	GetLocalResourceObject(String, Type, String)	Gets a page-level resource object based on the specified ResourceKey property, object type, and property name. (Inherited from TemplateControl.)
	GetPostBackClientEvent	**Obsolete.** Gets a reference that can be used in a client event to post back to the server for the specified control and with the specified

		event arguments. (Inherited from Page.)
	GetPostBackClientHyperlink	**Obsolete.** Gets a reference, with javascript: appended to the beginning of it, that can be used in a client event to post back to the server for the specified control and with the specified event arguments. (Inherited from Page.)
	GetPostBackEventReference(Control)	**Obsolete.** Returns a string that can be used in a client event to cause postback to the server. The reference string is defined by the specified Control object. (Inherited from Page.)

	GetPostBackEventReference(Control, String)	**Obsolete.** Returns a string that can be used in a client event to cause postback to the server. The reference string is defined by the specified control that handles the postback and a string argument of additional event information. (Inherited from Page.)
	GetRouteUrl(Object)	Gets the URL that corresponds to a set of route parameters. (Inherited from Control.)
	GetRouteUrl(RouteValueDictionary)	Gets the URL that corresponds to a set of route parameters. (Inherited from Control.)

	GetRouteUrl(String, Object)	Gets the URL that corresponds to a set of route parameters and a route name. (Inherited from Control.)
	GetRouteUrl(String, RouteValueDictionary)	Gets the URL that corresponds to a set of route parameters and a route name. (Inherited from Control.)
	GetType	Gets the Type of the current instance. (Inherited from Object.)
	GetTypeHashCode	Retrieves a hash code that is generated by Page objects that are generated at run time. This hash code is unique to the Page

		object's control hierarchy. (Inherited from Page.)
	GetUniqueIDRelativeTo	Returns the prefixed portion of the UniqueID property of the specified control. (Inherited from Control.)
	GetValidators	Returns a collection of control validators for a specified validation group. (Inherited from Page.)
	GetWrappedFileDependencies	Returns a list of physical file names that correspond to a list of virtual file locations. (Inherited from Page.)

	HasControls	Determines if the server control contains any child controls. (Inherited from Control.)
	HasEvents	Returns a value indicating whether events are registered for the control or any child controls. (Inherited from Control.)
	InitializeCulture	Sets the Culture and UICulture for the current thread of the page. (Inherited from Page.)
	InitOutputCache(OutputCacheParameters)	Initializes the output cache for the current page request based on an OutputCacheParamete

		rs object. (Inherited from Page.)
	InitOutputCache(Int32, String, String, OutputCacheLocation, String)	Initializes the output cache for the current page request. (Inherited from Page.)
	InitOutputCache(Int32, String, String, String, OutputCacheLocation, String)	Initializes the output cache for the current page request. (Inherited from Page.)
	IsClientScriptBlockRegistered	**Obsolete.** Determines whether the client script block with the specified key is registered with the page. (Inherited from Page.)
	IsLiteralContent	Determines if the server control holds only literal content.

		(Inherited from Control.)
	IsStartupScriptRegistered	**Obsolete.** Determines whether the client startup script is registered with the Page object. (Inherited from Page.)
	LoadControl(String)	Loads a Control object from a file based on a specified virtual path. (Inherited from TemplateControl.)
	LoadControl(Type, Object[])	Loads a Control object based on a specified type and constructor parameters. (Inherited from TemplateControl.)
	LoadControlState	Restores control-state information from a

		previous page request that was saved by the SaveControlState() method. (Inherited from Control.)
	LoadPageStateFromPersistenceMedium	Loads any saved view-state information to the Page object. (Inherited from Page.)
	LoadTemplate	Obtains an instance of the ITemplate interface from an external file. (Inherited from TemplateControl.)
	LoadViewState	Restores view-state information from a previous page request that was saved by the SaveViewState() method. (Inherited from

		Control.)
	MapPath	Retrieves the physical path that a virtual path, either absolute or relative, or an application-relative path maps to. (Inherited from Page.)
	MapPathSecure	Retrieves the physical path that a virtual path, either absolute or relative, maps to. (Inherited from Control.)
	MemberwiseClone	Creates a shallow copy of the current Object. (Inherited from Object.)
	OnAbortTransaction	Raises the AbortTransaction event.

	(Inherited from TemplateControl.)
OnBubbleEvent	Determines whether the event for the server control is passed up the page's UI server control hierarchy. (Inherited from Control.)
OnCommitTransaction	Raises the CommitTransaction event. (Inherited from TemplateControl.)
OnDataBinding	Raises the DataBinding event. (Inherited from Control.)
OnError	Raises the Error event. (Inherited from

		TemplateControl.)
	OnInit	Raises the Init event to initialize the page. (Inherited from Page.)
	OnInitComplete	Raises the InitComplete event after page initialization. (Inherited from Page.)
	OnLoad	Raises the Load event. (Inherited from Control.)
	OnLoadComplete	Raises the LoadComplete event at the end of the page load stage. (Inherited from Page.)
	OnPreInit	Raises the PreInit event at the beginning

	of page initialization. (Inherited from Page.)
OnPreLoad	Raises the PreLoad event after postback data is loaded into the page server controls but before the OnLoad(EventArgs) event. (Inherited from Page.)
OnPreRender	Raises the PreRender event. (Inherited from Control.)
OnPreRenderComplete	Raises the PreRenderComplete event after the OnPreRenderComplete (EventArgs) event and before the page is rendered. (Inherited from Page.)

OnSaveStateComplete	Raises the SaveStateComplete event after the page state has been saved to the persistence medium. (Inherited from Page.)
OnUnload	Raises the Unload event. (Inherited from Control.)
OpenFile	Gets a Stream used to read a file. (Inherited from Control.)
Page_Load	Page_Load.
ParseControl(String)	Parses an input string into a Control object on the Web Forms page or user control.

		(Inherited from TemplateControl.)
	ParseControl(String, Boolean)	Parses an input string into a Control object on the ASP.NET Web page or user control. (Inherited from TemplateControl.)
	ProcessRequest	Sets the intrinsic server objects of the Page object, such as the Context, Request, Response, and Application properties. (Inherited from Page.)
	RaiseBubbleEvent	Assigns any sources of the event and its information to the control's parent. (Inherited from Control.)

	RaisePostBackEvent	Notifies the server control that caused the postback that it should handle an incoming postback event. (Inherited from Page.)
	ReadStringResource	Reads a string resource. The ReadStringResource() method is not intended for use from within your code. (Inherited from TemplateControl.)
	RegisterArrayDeclaration	**Obsolete.** Declares a value that is declared as an ECMAScript array declaration when the page is rendered. (Inherited from Page.)
	RegisterAsyncTask	Registers a new

	asynchronous task with the page. (Inherited from Page.)
RegisterClientScriptBlock	**Obsolete.** Emits client-side script blocks to the response. (Inherited from Page.)
RegisterHiddenField	**Obsolete.** Allows server controls to automatically register a hidden field on the form. The field will be sent to the Page object when the HtmlForm server control is rendered. (Inherited from Page.)
RegisterOnSubmitStatement	**Obsolete.** Allows a page to access the client OnSubmit event. The script should be a

		function call to client code registered elsewhere. (Inherited from Page.)
	RegisterRequiresControlState	Registers a control as one whose control state must be persisted. (Inherited from Page.)
	RegisterRequiresPostBack	Registers a control as one that requires postback handling when the page is posted back to the server. (Inherited from Page.)
	RegisterRequiresRaiseEvent	Registers an ASP.NET server control as one requiring an event to be raised when the control is processed on the Page object.

		(Inherited from Page.)
	RegisterRequiresViewStateEncryption	Registers a control with the page as one requiring view-state encryption. (Inherited from Page.)
	RegisterStartupScript	**Obsolete.** Emits a client-side script block in the page response. (Inherited from Page.)
	RegisterViewStateHandler	Causes page view state to be persisted, if called. (Inherited from Page.)
	RemovedControl	Called after a child control is removed from the Controls collection of the Control object.

		(Inherited from Control.)
	Render	Initializes the HtmlTextWriter object and calls on the child controls of the Page to render. (Inherited from Page.)
	RenderChildren	Outputs the content of a server control's children to a provided HtmlTextWriter object, which writes the content to be rendered on the client. (Inherited from Control.)
	RenderControl(HtmlTextWriter)	Outputs server control content to a provided HtmlTextWriter object and stores tracing information about the control if tracing is

		enabled. (Inherited from Control.)
	RenderControl(HtmlTextWriter, ControlAdapter)	Outputs server control content to a provided HtmlTextWriter object using a provided ControlAdapter object. (Inherited from Control.)
	RequiresControlState	Determines whether the specified Control object is registered to participate in control state management. (Inherited from Page.)
	ResolveAdapter	Gets the control adapter responsible for rendering the specified control. (Inherited from Control.)

ResolveClientUrl	Gets a URL that can be used by the browser. (Inherited from Control.)
ResolveUrl	Converts a URL into one that is usable on the requesting client. (Inherited from Control.)
SaveControlState	Saves any server control state changes that have occurred since the time the page was posted back to the server. (Inherited from Control.)
SavePageStateToPersistenceMedium	Saves any view-state and control-state information for the page. (Inherited from Page.)

	SaveViewState	Saves any server control view-state changes that have occurred since the time the page was posted back to the server. (Inherited from Control.)
	SetDesignModeState	Sets design-time data for a control. (Inherited from Control.)
	SetFocus(Control)	Sets the browser focus to the specified control. (Inherited from Page.)
	SetFocus(String)	Sets the browser focus to the control with the specified identifier. (Inherited from Page.)

	SetRenderMethodDelegate	Assigns an event handler delegate to render the server control and its content into its parent control. (Inherited from Control.)
	SetStringResourcePointer	Sets a pointer to a string resource. The SetStringResourcePointer(Object, Int32) method is used by generated classes and is not intended for use from within your code. (Inherited from TemplateControl.)
	TestDeviceFilter	Returns a Boolean value indicating whether a device filter applies to the HTTP request. (Inherited from TemplateControl.)

	ToString	Returns a string that represents the current object. (Inherited from Object.)
	TrackViewState	Causes tracking of view-state changes to the server control so they can be stored in the server control's StateBag object. This object is accessible through the ViewState property. (Inherited from Control.)
	UnregisterRequiresControlState	Stops persistence of control state for the specified control. (Inherited from Page.)
	Validate()	Instructs any validation controls included on

		the page to validate their assigned information. (Inherited from Page.)
	Validate(String)	Instructs the validation controls in the specified validation group to validate their assigned information. (Inherited from Page.)
	VerifyRenderingInServerForm	Confirms that an HtmlForm control is rendered for the specified ASP.NET server control at run time. (Inherited from Page.)
	WriteUTF8ResourceString	Writes a resource string to an HtmlTextWriter control. The WriteUTF8ResourceString(HtmlTextWriter,

	Int32, Int32, Boolean) method is used by generated classes and is not intended for use from within your code. (Inherited from TemplateControl.)
XPath(String)	Evaluates an XPath data-binding expression. (Inherited from TemplateControl.)
XPath(String, IXmlNamespaceResolver)	Evaluates an XPath data-binding expression using the specified prefix and namespace mappings for namespace resolution. (Inherited from TemplateControl.)
XPath(String, String)	Evaluates an XPath data-binding

		expression using the specified format string to display the result. (Inherited from TemplateControl.)
	XPath(String, String, IXmlNamespaceResolver)	Evaluates an XPath data-binding expression using the specified prefix and namespace mappings for namespace resolution and the specified format string to display the result. (Inherited from TemplateControl.)
	XPathSelect(String)	Evaluates an XPath data-binding expression and returns a node collection that implements the IEnumerable interface. (Inherited from TemplateControl.)

	XPathSelect(String, IXmlNamespaceResolver)	Evaluates an XPath data-binding expression using the specified prefix and namespace mappings for namespace resolution and returns a node collection that implements the IEnumerable interface. (Inherited from TemplateControl.)

Top

Fields

	Name	Description
	ButtonCancel	ButtonCancel control.
	ButtonChangeVMPass	ButtonChangeVMPass control.
	ButtonCreateNewUser	ButtonCreateNewUser control.

	ButtonCreateVM	ButtonCreateVM control.
	CheckBoxCurrentCustomer	CheckBoxCurrentCustomer control.
	CustomerId	CustomerId control.
	CustomerNumberDL	CustomerNumberDL control.
	CustomerNumberTXT	CustomerNumberTXT control.
	form1	form1 control.
	LabelName	LabelName control.
	LabelPN	LabelPN control.
	LabelVM	LabelVM control.

	LabelVMPass	LabelVMPass control.
	ScriptManager1	ScriptManager1 control.
	TextboxName	TextboxName control.
	TextBoxNCustID	TextBoxNCustID control.
	TextBoxNewNumber	TextBoxNewNumber control.
	UpdatePanel1	UpdatePanel1 control.
	VoiceMailNumber	VoiceMailNumber control.
	VoiceMailPassTx	VoiceMailPassTx control.

Top

◢Properties

	Name	Description
	Adapter	Gets the browser-specific adapter for the control. (Inherited from Control.)
	Application	Gets the HttpApplicationState object for the current Web request. (Inherited from Page.)
	AppRelativeTemplateSourceDirectory	Gets or sets the application-relative virtual directory of the Page or UserControl object that contains this control.

		(Inherited from Control.)
	AppRelativeVirtualPath	Gets or sets the application-relative, virtual directory path to the file from which the control is parsed and compiled. (Inherited from TemplateControl.)
	AspCompatMode	Sets a value indicating whether the page can be executed on a single-threaded apartment (STA) thread. (Inherited from Page.)

	AsyncMode	Sets a value indicating whether the page is processed synchronously or asynchronously. (Inherited from Page.)
	AsyncTimeout	Gets or sets a value indicating the time-out interval used when processing asynchronous tasks. (Inherited from Page.)
	AutoHandlers	**Obsolete.** The AutoHandlers property has been deprecated in ASP.NET

		NET 2.0. It is used by generated classes and is not intended for use within your code. (Inherited from TemplateControl.)
	AutoPostBackControl	Gets or sets the control in the page that is used to perform postbacks. (Inherited from Page.)
	BindingContainer	Gets the control that contains this control's data binding. (Inherited from Control.)

	Name	Description
	Buffer	Sets a value indicating whether the page output is buffered. (Inherited from Page.)
	Cache	Gets the Cache object associated with the application in which the page resides. (Inherited from Page.)
	ChildControlsCreated	Gets a value that indicates whether the server control's child controls have been created. (Inherited from Control.)

	ClientID	Gets the control ID for HTML markup that is generated by ASP.NET. (Inherited from Control.)
	ClientIDMode	Gets or sets the algorithm that is used to generate the value of the ClientID property. (Inherited from Control.)
	ClientIDSeparator	Gets a character value representing the separator character used in the ClientID property. (Inherited from Control.)

	ClientQueryString	Gets the query string portion of the requested URL. (Inherited from Page.)
	ClientScript	Gets a ClientScriptManager object used to manage, register, and add script to the page. (Inherited from Page.)
	ClientTarget	Gets or sets a value that allows you to override automatic detection of browser capabilities and to specify how a page is rendered for particular

		browser clients. (Inherited from Page.)
	CodePage	Sets the code page identifier for the current Page. (Inherited from Page.)
	ContentType	Sets the HTTP MIME type for the HttpResponse object associated with the page. (Inherited from Page.)
	Context	Gets the HttpContext object associated with the page. (Inherited from

		Page.)
	Controls	Gets a ControlCollection object that represents the child controls for a specified server control in the UI hierarchy. (Inherited from Control.)
	Culture	Sets the culture ID for the Thread object associated with the page. (Inherited from Page.)
	DataItemContainer	Gets a reference to the naming container if the naming container implements

		IDataItemContainer. (Inherited from Control.)
	DataKeysContainer	Gets a reference to the naming container if the naming container implements IDataKeysControl. (Inherited from Control.)
	DesignMode	Gets a value indicating whether a control is being used on a design surface. (Inherited from Control.)

	EnableEventValidation	Gets or sets a value indicating whether the page validates postback and callback events. (Inherited from Page.)
	EnableTheming	Gets or sets a Boolean value indicating whether themes apply to the control that is derived from the TemplateControl class. (Inherited from TemplateControl.)
	EnableViewState	Gets or sets a value indicating whether the page maintains its view state,

		and the view state of any server controls it contains, when the current page request ends. (Inherited from Page.)
	EnableViewStateMac	Gets or sets a value indicating whether ASP.NET should check message authentication codes (MAC) in the page's view state when the page is posted back from the client. (Inherited from Page.)
	ErrorPage	Gets or sets the error page to which the

		requesting browser is redirected in the event of an unhandled page exception. (Inherited from Page.)
	Events	Gets a list of event handler delegates for the control. This property is read-only. (Inherited from Control.)
	FileDependencies	**Obsolete.** Sets an array of files that the current HttpResponse object is dependent upon. (Inherited from

		Page.)
	Form	Gets the HTML form for the page. (Inherited from Page.)
	HasChildViewState	Gets a value indicating whether the current server control's child controls have any saved view-state settings. (Inherited from Control.)
	Header	Gets the document header for the page if the head element is defined with a runat=server in

		the page declaration. (Inherited from Page.)
	ID	Gets or sets an identifier for a particular instance of the Page class. (Inherited from Page.)
	IdSeparator	Gets the character used to separate control identifiers when building a unique ID for a control on a page. (Inherited from Page.)
	IsAsync	Gets a value indicating

		whether the page is processed asynchronously. (Inherited from Page.)
	IsCallback	Gets a value that indicates whether the page request is the result of a callback. (Inherited from Page.)
	IsChildControlStateCleared	Gets a value indicating whether controls contained within this control have control state. (Inherited from Control.)

	IsCrossPagePostBack	Gets a value indicating whether the page is involved in a cross-page postback. (Inherited from Page.)
	IsPostBack	Gets a value that indicates whether the page is being rendered for the first time or is being loaded in response to a postback. (Inherited from Page.)
	IsPostBackEventControlRegistered	Gets a value that indicates whether the control in the page that performs

		postbacks has been registered. (Inherited from Page.)
	IsReusable	Gets a value indicating whether the Page object can be reused. (Inherited from Page.)
	IsTrackingViewState	Gets a value that indicates whether the server control is saving changes to its view state. (Inherited from Control.)
	IsValid	Gets a value indicating whether page validation

		succeeded. (Inherited from Page.)
	IsViewStateEnabled	Gets a value indicating whether view state is enabled for this control. (Inherited from Control.)
	Items	Gets a list of objects stored in the page context. (Inherited from Page.)
	LCID	Sets the locale identifier for the Thread object associated with the page. (Inherited from

		Page.)
	LoadViewStateByID	Gets a value indicating whether the control participates in loading its view state by ID instead of index. (Inherited from Control.)
	MaintainScrollPositionOnPostBack	Gets or sets a value indicating whether to return the user to the same position in the client browser after postback. This property replaces the obsolete SmartNavigation property. (Inherited from

		Page.)
	Master	Gets the master page that determines the overall look of the page. (Inherited from Page.)
	MasterPageFile	Gets or sets the virtual path of the master page. (Inherited from Page.)
	MaxPageStateFieldLength	Gets or sets the maximum length for the page's state field. (Inherited from Page.)
	MetaDescription	Gets or sets the content of the

		"description" meta element. (Inherited from Page.)
	MetaKeywords	Gets or sets the content of the "keywords" meta element. (Inherited from Page.)
	NamingContainer	Gets a reference to the server control's naming container, which creates a unique namespace for differentiating between server controls with the same ID property value. (Inherited from Control.)

	Page	Gets a reference to the Page instance that contains the server control. (Inherited from Control.)
	PageAdapter	Gets the adapter that renders the page for the specific requesting browser. (Inherited from Page.)
	PageStatePersister	Gets the PageStatePersister object associated with the page. (Inherited from Page.)

	Parent	Gets a reference to the server control's parent control in the page control hierarchy. (Inherited from Control.)
	PreviousPage	Gets the page that transferred control to the current page. (Inherited from Page.)
	RenderingCompatibility	Gets a value that specifies the ASP.NET version that rendered HTML will be compatible with. (Inherited from Control.)

	Request	Gets the HttpRequest object for the requested page. (Inherited from Page.)
	Response	Gets the HttpResponse object associated with the Page object. This object allows you to send HTTP response data to a client and contains information about that response. (Inherited from Page.)
	ResponseEncoding	Sets the encoding language for the current

		HttpResponse object. (Inherited from Page.)
	RouteData	Gets the RouteData value of the current RequestContext instance. (Inherited from Page.)
	Server	Gets the Server object, which is an instance of the HttpServerUtility class. (Inherited from Page.)
	Session	Gets the current Session object provided by

		ASP.NET. (Inherited from Page.)
	Site	Gets information about the container that hosts the current control when rendered on a design surface. (Inherited from Control.)
	SkinID	Gets or sets the skin to apply to the control. (Inherited from Control.)
	SmartNavigation	**Obsolete.** Gets or sets a value indicating whether smart navigation is

		enabled. This property is deprecated. (Inherited from Page.)
	StyleSheetTheme	Gets or sets the name of the theme that is applied to the page early in the page life cycle. (Inherited from Page.)
	SupportAutoEvents	Gets a value indicating whether the TemplateControl control supports automatic events. (Inherited from TemplateControl.)

	TemplateControl	Gets or sets a reference to the template that contains this control. (Inherited from Control.)
	TemplateSourceDirectory	Gets the virtual directory of the Page or UserControl that contains the current server control. (Inherited from Control.)
	Theme	Gets or sets the name of the page theme. (Inherited from Page.)
	Title	Gets or sets the

		title for the page. (Inherited from Page.)
	Trace	Gets the TraceContext object for the current Web request. (Inherited from Page.)
	TraceEnabled	Sets a value indicating whether tracing is enabled for the Page object. (Inherited from Page.)
	TraceModeValue	Sets the mode in which trace statements are displayed on the page.

		(Inherited from Page.)
	TransactionMode	Sets the level of transaction support for the page. (Inherited from Page.)
	UICulture	Sets the user interface (UI) ID for the Thread object associated with the page. (Inherited from Page.)
	UniqueFilePathSuffix	Gets a unique suffix to append to the file path for caching browsers. (Inherited from Page.)

	UniqueID	Gets the unique, hierarchically qualified identifier for the server control. (Inherited from Control.)
	User	Gets information about the user making the page request. (Inherited from Page.)
	Validators	Gets a collection of all validation controls contained on the requested page. (Inherited from Page.)
	ViewState	Gets a dictionary of state

		information that allows you to save and restore the view state of a server control across multiple requests for the same page. (Inherited from Control.)
	ViewStateEncryptionMode	Gets or sets the encryption mode of the view state. (Inherited from Page.)
	ViewStateIgnoresCase	Gets a value that indicates whether the StateBag object is case-insensitive. (Inherited from Control.)

	ViewStateMode	Gets or sets the view-state mode of this control. (Inherited from Control.)
	ViewStateUserKey	Assigns an identifier to an individual user in the view-state variable associated with the current page. (Inherited from Page.)
	Visible	Gets or sets a value indicating whether the Page object is rendered. (Inherited from Page.)

Top

Events

	Name	Description
	AbortTransaction	Occurs when a user ends a transaction. (Inherited from TemplateControl.)
	CommitTransaction	Occurs when a transaction completes. (Inherited from TemplateControl.)
	DataBinding	Occurs when the server control binds to a data source. (Inherited from Control.)
	Disposed	Occurs when a server control is released from memory, which is the last stage of the server control

		lifecycle when an ASP.NET page is requested. (Inherited from Control.)
	Error	Occurs when an unhandled exception is thrown. (Inherited from TemplateControl.)
	Init	Occurs when the server control is initialized, which is the first step in its lifecycle. (Inherited from Control.)
	InitComplete	Occurs when page initialization is complete. (Inherited from Page.)

	Load	Occurs when the server control is loaded into the Page object. (Inherited from Control.)
	LoadComplete	Occurs at the end of the load stage of the page's life cycle. (Inherited from Page.)
	PreInit	Occurs at the beginning of page initialization. (Inherited from Page.)
	PreLoad	Occurs before the page Load event. (Inherited from Page.)

	PreRender	Occurs after the Control object is loaded but prior to rendering. (Inherited from Control.)
	PreRenderComplete	Occurs before the page content is rendered. (Inherited from Page.)
	SaveStateComplete	Occurs after the page has completed saving all view state and control state information for the page and controls on the page. (Inherited from Page.)
	Unload	Occurs when the server control is

unloaded from memory.

(Inherited from Control.)

Top

Remarks

Simply half of the full class one half being the web page and the other being the code behind

Remarks

Simply half of the full class one half being the web page and the other being the code behind

Default Constructor

Initializes a new instance of the Default class

Namespace: AsteriskManager

Assembly: AsteriskManager (in AsteriskManager.dll) Version: 1.0.0.0 (1.0.0.0)

◢Syntax

C#

VB

C++

F#

Copy

```
public Default()
```

Default Fields

The Default type exposes the following members.

◢Fields

	Name	Description
	ButtonCancel	ButtonCancel control.
	ButtonChangeVMPass	ButtonChangeVMPass control.
	ButtonCreateNewUser	ButtonCreateNewUser control.
	ButtonCreateVM	ButtonCreateVM control.
	CheckBoxCurrentCustomer	CheckBoxCurrentCustomer control.

	CustomerId	CustomerId control.
	CustomerNumberDL	CustomerNumberDL control.
	CustomerNumberTXT	CustomerNumberTXT control.
	form1	form1 control.
	LabelName	LabelName control.
	LabelPN	LabelPN control.
	LabelVM	LabelVM control.
	LabelVMPass	LabelVMPass control.
	ScriptManager1	ScriptManager1 control.
	TextboxName	TextboxName control.

	TextBoxNCustID	TextBoxNCustID control.
	TextBoxNewNumber	TextBoxNewNumber control.
	UpdatePanel1	UpdatePanel1 control.
	VoiceMailNumber	VoiceMailNumber control.
	VoiceMailPassTx	VoiceMailPassTx control.

Default. ButtonCancel Field

ButtonCancel control.

Namespace: AsteriskManager

Assembly: AsteriskManager (in AsteriskManager.dll) Version: 1.0.0.0 (1.0.0.0)

◢Syntax

C#

VB

C++

F#

Copy

```
protected Button ButtonCancel
```

Type: Button

◢Remarks

Auto-generated field. To modify move field declaration from designer file to code-behind file.

Default.ButtonChangeVMPass Field

ButtonChangeVMPass control.
Namespace: AsteriskManager
Assembly: AsteriskManager (in AsteriskManager.dll) Version: 1.0.0.0 (1.0.0.0)

◢Syntax

C#

VB

C++

F#

Copy

```
protected Button ButtonChangeVMPass
```

Type: Button

◢Remarks

Auto-generated field. To modify move field declaration from designer file to code-behind file.

Default. ButtonCreateNewUser Field

ButtonCreateNewUser control.
Namespace: AsteriskManager
Assembly: AsteriskManager (in AsteriskManager.dll) Version: 1.0.0.0 (1.0.0.0)

◢Syntax

C#

VB

C++

F#

Copy

```
protected Button ButtonCreateNewUser
```

Type: Button

◢Remarks

Auto-generated field. To modify move field declaration from designer file to code-behind file.

Default. ButtonCreateVM Field

ButtonCreateVM control.
Namespace: AsteriskManager
Assembly: AsteriskManager (in AsteriskManager.dll) Version: 1.0.0.0 (1.0.0.0)

◢Syntax

C#

VB

C++

F#

Copy

```
protected Button ButtonCreateVM
```

Type: Button

Remarks

Auto-generated field. To modify move field declaration from designer file to code-behind file.

Default. CheckBoxCurrentCustomer Field

CheckBoxCurrentCustomer control.
Namespace: AsteriskManager
Assembly: AsteriskManager (in AsteriskManager.dll) Version: 1.0.0.0 (1.0.0.0)

Syntax

C#

VB

C++

F#

Copy

```
protected CheckBox CheckBoxCurrentCustomer
```

Type: CheckBox

Remarks

Auto-generated field. To modify move field declaration from designer file to code-behind file.

Default. CustomerId Field

CustomerId control.

Namespace: AsteriskManager

Assembly: AsteriskManager (in AsteriskManager.dll) Version: 1.0.0.0 (1.0.0.0)

Syntax

C#

VB

C++

F#

Copy

```
protected DropDownList CustomerId
```

Type: DropDownList

◢Remarks

Auto-generated field. To modify move field declaration from designer file to code-behind file.

Default.CustomerNumberDL Field

CustomerNumberDL control.
Namespace: AsteriskManager
Assembly: AsteriskManager (in AsteriskManager.dll) Version: 1.0.0.0 (1.0.0.0)

◢Syntax

C#

VB

C++

F#

Copy

```
protected DropDownList CustomerNumberDL
```

Type: DropDownList

◢Remarks

Auto-generated field. To modify move field declaration from designer file to code-behind file.

Default.CustomerNumberTXT Field

CustomerNumberTXT control.
Namespace: AsteriskManager
Assembly: AsteriskManager (in AsteriskManager.dll) Version: 1.0.0.0 (1.0.0.0)

◢Syntax

C#

VB

C++

F#

Copy

```
protected TextBox CustomerNumberTXT
```

Type: TextBox

Remarks

Auto-generated field. To modify move field declaration from designer file to code-behind file.

Default. form1 Field

form1 control.
Namespace: AsteriskManager
Assembly: AsteriskManager (in AsteriskManager.dll) Version: 1.0.0.0 (1.0.0.0)

Syntax

C#

VB

C++

F#

Copy

```
protected HtmlForm form1
```

Type: HtmlForm

Remarks

Auto-generated field. To modify move field declaration from designer file to code-behind file.

Default. LabelName Field

LabelName control.

Namespace: AsteriskManager

Assembly: AsteriskManager (in AsteriskManager.dll) Version: 1.0.0.0 (1.0.0.0)

Syntax

C#

VB

C++

F#

Copy

```
protected Label LabelName
```

Type: Label

Remarks

Auto-generated field. To modify move field declaration from designer file to code-behind file.

Default. LabelPN Field

LabelPN control.

Namespace: AsteriskManager

Assembly: AsteriskManager (in AsteriskManager.dll) Version: 1.0.0.0 (1.0.0.0)

Syntax

C#

VB

C++

F#

Copy

```
protected Label LabelPN
```

Type: Label

Remarks

Auto-generated field. To modify move field declaration from designer file to code-behind file.

Default. LabelVM Field

LabelVM control.

Namespace: AsteriskManager

Assembly: AsteriskManager (in AsteriskManager.dll) Version: 1.0.0.0 (1.0.0.0)

◢Syntax

C#

VB

C++

F#

Copy

```
protected Label LabelVM
```

Type: Label

◢Remarks

Auto-generated field. To modify move field declaration from designer file to code-behind file.

Default. LabelVMPass Field

LabelVMPass control.
Namespace: AsteriskManager
Assembly: AsteriskManager (in AsteriskManager.dll) Version: 1.0.0.0 (1.0.0.0)

◢Syntax

C#

VB

C++

F#

Copy

```
protected Label LabelVMPass
```

Type: Label

◢Remarks

Auto-generated field. To modify move field declaration from designer file to code-behind file.

Default. ScriptManager1 Field

ScriptManager1 control.
Namespace: AsteriskManager
Assembly: AsteriskManager (in AsteriskManager.dll) Version: 1.0.0.0 (1.0.0.0)

◢Syntax

C#

VB

C++

F#

Copy

```
protected ScriptManager ScriptManager1
```

Type: ScriptManager

Remarks

Auto-generated field. To modify move field declaration from designer file to code-behind file.

Default. TextboxName Field

TextboxName control.
Namespace: AsteriskManager
Assembly: AsteriskManager (in AsteriskManager.dll) Version: 1.0.0.0 (1.0.0.0)

Syntax

C#

VB

C++

F#

Copy

```
protected TextBox TextboxName
```

Type: TextBox

Remarks

Auto-generated field. To modify move field declaration from designer file to code-behind file.

Default.TextBoxNCustID Field

TextBoxNCustID control.
Namespace: AsteriskManager
Assembly: AsteriskManager (in AsteriskManager.dll) Version: 1.0.0.0 (1.0.0.0)

Syntax

C#

VB

C++

F#

Copy

```
protected TextBox TextBoxNCustID
```

Type: TextBox

Remarks

Auto-generated field. To modify move field declaration from designer file to code-behind file.

Default. TextBoxNewNumber Field

TextBoxNewNumber control.

Namespace: AsteriskManager

Assembly: AsteriskManager (in AsteriskManager.dll) Version: 1.0.0.0 (1.0.0.0)

Syntax

C#

VB

C++

F#

Copy

```
protected TextBox TextBoxNewNumber
```

Type: TextBox

Remarks

Auto-generated field. To modify move field declaration from designer file to code-behind file.

Default. UpdatePanel1 Field

UpdatePanel1 control.
Namespace: AsteriskManager
Assembly: AsteriskManager (in AsteriskManager.dll) Version: 1.0.0.0 (1.0.0.0)

Syntax

C#

VB

C++

F#

Copy

```
protected UpdatePanel UpdatePanel1
```

Type: UpdatePanel

Remarks

Auto-generated field. To modify move field declaration from designer file to code-behind file.

Default. VoiceMailNumber Field

VoiceMailNumber control.

Namespace: AsteriskManager

Assembly: AsteriskManager (in AsteriskManager.dll) Version: 1.0.0.0 (1.0.0.0)

Syntax

C#

VB

C++

F#

Copy

```
protected TextBox VoiceMailNumber
```

Type: TextBox

Remarks

Auto-generated field. To modify move field declaration from designer file to code-behind file.

Default.VoiceMailPassTx Field

VoiceMailPassTx control.

Namespace: AsteriskManager

Assembly: AsteriskManager (in AsteriskManager.dll) Version: 1.0.0.0 (1.0.0.0)

Syntax

C#

VB

C++

F#

Copy

```
protected TextBox VoiceMailPassTx
```

Type: TextBox

Remarks

Auto-generated field. To modify move field declaration from designer file to code-behind file.

Default Methods

The Default type exposes the following members.

Methods

Name	Description
AddContentTemplate	Called during page initialization to create a collection of content (from content controls) that is handed to a master page, if the current page or master page refers to a master page. (Inherited from Page.)
AddedControl	Called after a child control is added to the Controls collection of the Control object. (Inherited from Control.)
AddOnPreRenderComplet	Registers beginning

	eAsync(BeginEventHandler, EndEventHandler)	and ending event handler delegates that do not require state information for an asynchronous page. (Inherited from Page.)
	AddOnPreRenderCompleteAsync(BeginEventHandler, EndEventHandler, Object)	Registers beginning and ending event handler delegates for an asynchronous page. (Inherited from Page.)
	AddParsedSubObject	Notifies the server control that an element, either XML or HTML, was parsed, and adds the element to the server control's ControlCollection object. (Inherited from Control.)
	AddWrappedFileDepende	Adds a list of dependent files that

	ncies	make up the current page. This method is used internally by the ASP.NET page framework and is not intended to be used directly from your code. (Inherited from Page.)
	ApplyStyleSheetSkin	Applies the style properties defined in the page style sheet to the control. (Inherited from Control.)
	AspCompatBeginProcessRequest	Initiates a request for Active Server Page (ASP) resources. This method is provided for compatibility with legacy ASP applications. (Inherited from Page.)

AspCompatEndProcessRequest	Terminates a request for Active Server Page (ASP) resources. This method is provided for compatibility with legacy ASP applications. (Inherited from Page.)
AsyncPageBeginProcessRequest	Begins processing an asynchronous page request. (Inherited from Page.)
AsyncPageEndProcessRequest	Ends processing an asynchronous page request. (Inherited from Page.)
BuildProfileTree	Gathers information about the server control and delivers it to the Trace property to be displayed when tracing is enabled for

		the page. (Inherited from Control.)
	ButtonCancel_Click	ButtonCancel_Click.
	ButtonChangeVMPass_Click	ButtonChangeVMPass_Click.
	ButtonCreateNewUser_Click	ButtonCreateNewUser_Click.
	ButtonCreateVM_Click	ButtonCreateVM_Click.
	CheckBoxCurrentCustomer_CheckedChanged	CheckBoxCurrentCustomer_CheckedChanged.
	ClearCachedClientID	Sets the cached ClientID value to null. (Inherited from Control.)
	ClearChildControlState	Deletes the control-

		state information for the server control's child controls. (Inherited from Control.)
	ClearChildState	Deletes the view-state and control-state information for all the server control's child controls. (Inherited from Control.)
	ClearChildViewState	Deletes the view-state information for all the server control's child controls. (Inherited from Control.)
	ClearEffectiveClientIDMode	Sets the ClientIDMode property of the current control instance and of any child controls to

		Inherit. (Inherited from Control.)
	Construct	Performs design-time logic. (Inherited from TemplateControl.)
	CreateChildControls	Called by the ASP.NET page framework to notify server controls that use composition-based implementation to create any child controls they contain in preparation for posting back or rendering. (Inherited from Control.)
	CreateControlCollection	Creates a new ControlCollection object to hold the child controls (both

		literal and server) of the server control. (Inherited from Control.)
	CreateHtmlTextWriter	Creates an HtmlTextWriter object to render the page's content. (Inherited from Page.)
	CreateResourceBasedLiteralControl	Accesses literal strings stored in a resource. The CreateResourceBasedLiteralControl(Int32, Int32, Boolean) method is not intended for use from within your code. (Inherited from TemplateControl.)
	CustomerId_SelectedIndexChanged	CustomerId_SelectedIndexChanged.

	CustomerNumberDL_SelectedIndexChanged	CustomerNumberDL_SelectedIndexChanged.
	CustomerNumberTXT_TextChanged	CustomerNumberTXT_TextChanged.
	DataBind()	Binds a data source to the invoked server control and all its child controls. (Inherited from Control.)
	DataBind(Boolean)	Binds a data source to the invoked server control and all its child controls with an option to raise the DataBinding event. (Inherited from Control.)
	DataBindChildren	Binds a data source to the server control's child controls.

		(Inherited from Control.)
	DesignerInitialize	Performs any initialization of the instance of the Page class that is required by RAD designers. This method is used only at design time. (Inherited from Page.)
	DeterminePostBackMode	Returns a NameValueCollection of data posted back to the page using either a POST or a GET command. (Inherited from Page.)
	Dispose	Enables a server control to perform final clean up before it is released from memory. (Inherited from

		Control.)
	EnsureChildControls	Determines whether the server control contains child controls. If it does not, it creates child controls. (Inherited from Control.)
	EnsureID	Creates an identifier for controls that do not have an identifier assigned. (Inherited from Control.)
	Equals	Determines whether the specified Object is equal to the current Object. (Inherited from Object.)
	Eval(String)	Evaluates a data-

	binding expression. (Inherited from TemplateControl.)
Eval(String, String)	Evaluates a data-binding expression using the specified format string to display the result. (Inherited from TemplateControl.)
ExecuteRegisteredAsyncTasks	Starts the execution of an asynchronous task. (Inherited from Page.)
fill_Drop_Down	fill_Drop_Down.
Finalize	Allows an object to try to free resources and perform other cleanup operations before it is reclaimed by garbage collection. (Inherited from

		Object.)
	FindControl(String)	Searches the page naming container for a server control with the specified identifier. (Inherited from Page.)
	FindControl(String, Int32)	Searches the current naming container for a server control with the specified *id* and an integer, specified in the *pathOffset* parameter, which aids in the search. You should not override this version of the FindControl() method. (Inherited from Control.)
	Focus	Sets input focus to a control. (Inherited from

		Control.)
	FrameworkInitialize	Initializes the control tree during page generation based on the declarative nature of the page. (Inherited from Page.)
	GetDataItem	Gets the data item at the top of the data-binding context stack. (Inherited from Page.)
	GetDesignModeState	Gets design-time data for a control. (Inherited from Control.)
	GetGlobalResourceObject(String, String)	Gets an application-level resource object based on the specified ClassKey and ResourceKey properties.

		(Inherited from TemplateControl.)
	GetGlobalResourceObject(String, String, Type, String)	Gets an application-level resource object based on the specified ClassKey and ResourceKey properties, object type, and property name of the resource. (Inherited from TemplateControl.)
	GetHashCode	Serves as a hash function for a particular type. (Inherited from Object.)
	GetLocalResourceObject(String)	Gets a page-level resource object based on the specified ResourceKey property. (Inherited from

		TemplateControl.)
	GetLocalResourceObject(String, Type, String)	Gets a page-level resource object based on the specified ResourceKey property, object type, and property name. (Inherited from TemplateControl.)
	GetPostBackClientEvent	**Obsolete.** Gets a reference that can be used in a client event to post back to the server for the specified control and with the specified event arguments. (Inherited from Page.)
	GetPostBackClientHyperlink	**Obsolete.** Gets a reference, with javascript: appended to the beginning of it,

		that can be used in a client event to post back to the server for the specified control and with the specified event arguments. (Inherited from Page.)
	GetPostBackEventReference(Control)	**Obsolete.** Returns a string that can be used in a client event to cause postback to the server. The reference string is defined by the specified Control object. (Inherited from Page.)
	GetPostBackEventReference(Control, String)	**Obsolete.** Returns a string that can be used in a client event to cause postback to the server. The reference string is defined by the

		specified control that handles the postback and a string argument of additional event information. (Inherited from Page.)
	GetRouteUrl(Object)	Gets the URL that corresponds to a set of route parameters. (Inherited from Control.)
	GetRouteUrl(RouteValueDictionary)	Gets the URL that corresponds to a set of route parameters. (Inherited from Control.)
	GetRouteUrl(String, Object)	Gets the URL that corresponds to a set of route parameters and a route name. (Inherited from Control.)

	GetRouteUrl(String, RouteValueDictionary)	Gets the URL that corresponds to a set of route parameters and a route name. (Inherited from Control.)
	GetType	Gets the Type of the current instance. (Inherited from Object.)
	GetTypeHashCode	Retrieves a hash code that is generated by Page objects that are generated at run time. This hash code is unique to the Page object's control hierarchy. (Inherited from Page.)
	GetUniqueIDRelativeTo	Returns the prefixed portion of the UniqueID property of

		the specified control. (Inherited from Control.)
	GetValidators	Returns a collection of control validators for a specified validation group. (Inherited from Page.)
	GetWrappedFileDependencies	Returns a list of physical file names that correspond to a list of virtual file locations. (Inherited from Page.)
	HasControls	Determines if the server control contains any child controls. (Inherited from Control.)
	HasEvents	Returns a value indicating whether

	events are registered for the control or any child controls. (Inherited from Control.)
InitializeCulture	Sets the Culture and UICulture for the current thread of the page. (Inherited from Page.)
InitOutputCache(OutputCacheParameters)	Initializes the output cache for the current page request based on an OutputCacheParameters object. (Inherited from Page.)
InitOutputCache(Int32, String, String, OutputCacheLocation, String)	Initializes the output cache for the current page request. (Inherited from Page.)

	InitOutputCache(Int32, String, String, String, OutputCacheLocation, String)	Initializes the output cache for the current page request. (Inherited from Page.)
	IsClientScriptBlockRegistered	**Obsolete.** Determines whether the client script block with the specified key is registered with the page. (Inherited from Page.)
	IsLiteralContent	Determines if the server control holds only literal content. (Inherited from Control.)
	IsStartupScriptRegistered	**Obsolete.** Determines whether the client startup script is registered with the Page object.

		(Inherited from Page.)
	LoadControl(String)	Loads a Control object from a file based on a specified virtual path. (Inherited from TemplateControl.)
	LoadControl(Type, Object[])	Loads a Control object based on a specified type and constructor parameters. (Inherited from TemplateControl.)
	LoadControlState	Restores control-state information from a previous page request that was saved by the SaveControlState() method. (Inherited from Control.)
	LoadPageStateFromPersist	Loads any saved view-

enceMedium	state information to the Page object. (Inherited from Page.)
LoadTemplate	Obtains an instance of the ITemplate interface from an external file. (Inherited from TemplateControl.)
LoadViewState	Restores view-state information from a previous page request that was saved by the SaveViewState() method. (Inherited from Control.)
MapPath	Retrieves the physical path that a virtual path, either absolute or relative, or an application-relative path maps to.

		(Inherited from Page.)
	MapPathSecure	Retrieves the physical path that a virtual path, either absolute or relative, maps to. (Inherited from Control.)
	MemberwiseClone	Creates a shallow copy of the current Object. (Inherited from Object.)
	OnAbortTransaction	Raises the AbortTransaction event. (Inherited from TemplateControl.)
	OnBubbleEvent	Determines whether the event for the server control is passed up the page's UI server control

	hierarchy. (Inherited from Control.)
OnCommitTransaction	Raises the CommitTransaction event. (Inherited from TemplateControl.)
OnDataBinding	Raises the DataBinding event. (Inherited from Control.)
OnError	Raises the Error event. (Inherited from TemplateControl.)
OnInit	Raises the Init event to initialize the page. (Inherited from Page.)

	OnInitComplete	Raises the InitComplete event after page initialization. (Inherited from Page.)
	OnLoad	Raises the Load event. (Inherited from Control.)
	OnLoadComplete	Raises the LoadComplete event at the end of the page load stage. (Inherited from Page.)
	OnPreInit	Raises the PreInit event at the beginning of page initialization. (Inherited from Page.)
	OnPreLoad	Raises the PreLoad event after postback data is loaded into the

		page server controls but before the OnLoad(EventArgs) event. (Inherited from Page.)
	OnPreRender	Raises the PreRender event. (Inherited from Control.)
	OnPreRenderComplete	Raises the PreRenderComplete event after the OnPreRenderComplete (EventArgs) event and before the page is rendered. (Inherited from Page.)
	OnSaveStateComplete	Raises the SaveStateComplete event after the page state has been saved to the persistence

		medium. (Inherited from Page.)
	OnUnload	Raises the Unload event. (Inherited from Control.)
	OpenFile	Gets a Stream used to read a file. (Inherited from Control.)
	Page_Load	Page_Load.
	ParseControl(String)	Parses an input string into a Control object on the Web Forms page or user control. (Inherited from TemplateControl.)
	ParseControl(String, Boolean)	Parses an input string into a Control object

		on the ASP.NET Web page or user control. (Inherited from TemplateControl.)
	ProcessRequest	Sets the intrinsic server objects of the Page object, such as the Context, Request, Response, and Application properties. (Inherited from Page.)
	RaiseBubbleEvent	Assigns any sources of the event and its information to the control's parent. (Inherited from Control.)
	RaisePostBackEvent	Notifies the server control that caused the postback that it should handle an incoming postback event.

		(Inherited from Page.)
	ReadStringResource	Reads a string resource. The ReadStringResource() method is not intended for use from within your code. (Inherited from TemplateControl.)
	RegisterArrayDeclaration	**Obsolete.** Declares a value that is declared as an ECMAScript array declaration when the page is rendered. (Inherited from Page.)
	RegisterAsyncTask	Registers a new asynchronous task with the page. (Inherited from Page.)

	RegisterClientScriptBlock	**Obsolete.** Emits client-side script blocks to the response. (Inherited from Page.)
	RegisterHiddenField	**Obsolete.** Allows server controls to automatically register a hidden field on the form. The field will be sent to the Page object when the HtmlForm server control is rendered. (Inherited from Page.)
	RegisterOnSubmitStatement	**Obsolete.** Allows a page to access the client OnSubmit event. The script should be a function call to client code registered elsewhere.

		(Inherited from Page.)
	RegisterRequiresControlState	Registers a control as one whose control state must be persisted. (Inherited from Page.)
	RegisterRequiresPostBack	Registers a control as one that requires postback handling when the page is posted back to the server. (Inherited from Page.)
	RegisterRequiresRaiseEvent	Registers an ASP.NET server control as one requiring an event to be raised when the control is processed on the Page object. (Inherited from Page.)
	RegisterRequiresViewState	Registers a control

Encryption	with the page as one requiring view-state encryption. (Inherited from Page.)
RegisterStartupScript	**Obsolete.** Emits a client-side script block in the page response. (Inherited from Page.)
RegisterViewStateHandler	Causes page view state to be persisted, if called. (Inherited from Page.)
RemovedControl	Called after a child control is removed from the Controls collection of the Control object. (Inherited from Control.)

	Render	Initializes the HtmlTextWriter object and calls on the child controls of the Page to render. (Inherited from Page.)
	RenderChildren	Outputs the content of a server control's children to a provided HtmlTextWriter object, which writes the content to be rendered on the client. (Inherited from Control.)
	RenderControl(HtmlTextWriter)	Outputs server control content to a provided HtmlTextWriter object and stores tracing information about the control if tracing is enabled. (Inherited from Control.)

	RenderControl(HtmlTextWriter, ControlAdapter)	Outputs server control content to a provided HtmlTextWriter object using a provided ControlAdapter object. (Inherited from Control.)
	RequiresControlState	Determines whether the specified Control object is registered to participate in control state management. (Inherited from Page.)
	ResolveAdapter	Gets the control adapter responsible for rendering the specified control. (Inherited from Control.)
	ResolveClientUrl	Gets a URL that can be used by the browser. (Inherited from

		Control.)
	ResolveUrl	Converts a URL into one that is usable on the requesting client. (Inherited from Control.)
	SaveControlState	Saves any server control state changes that have occurred since the time the page was posted back to the server. (Inherited from Control.)
	SavePageStateToPersistenceMedium	Saves any view-state and control-state information for the page. (Inherited from Page.)
	SaveViewState	Saves any server control view-state

		changes that have occurred since the time the page was posted back to the server. (Inherited from Control.)
	SetDesignModeState	Sets design-time data for a control. (Inherited from Control.)
	SetFocus(Control)	Sets the browser focus to the specified control. (Inherited from Page.)
	SetFocus(String)	Sets the browser focus to the control with the specified identifier. (Inherited from Page.)
	SetRenderMethodDelegat	Assigns an event handler delegate to

	e	render the server control and its content into its parent control. (Inherited from Control.)
	SetStringResourcePointer	Sets a pointer to a string resource. The SetStringResourcePointer(Object, Int32) method is used by generated classes and is not intended for use from within your code. (Inherited from TemplateControl.)
	TestDeviceFilter	Returns a Boolean value indicating whether a device filter applies to the HTTP request. (Inherited from TemplateControl.)

	ToString	Returns a string that represents the current object. (Inherited from Object.)
	TrackViewState	Causes tracking of view-state changes to the server control so they can be stored in the server control's StateBag object. This object is accessible through the ViewState property. (Inherited from Control.)
	UnregisterRequiresControlState	Stops persistence of control state for the specified control. (Inherited from Page.)
	Validate()	Instructs any validation controls included on

		the page to validate their assigned information. (Inherited from Page.)
	Validate(String)	Instructs the validation controls in the specified validation group to validate their assigned information. (Inherited from Page.)
	VerifyRenderingInServerForm	Confirms that an HtmlForm control is rendered for the specified ASP.NET server control at run time. (Inherited from Page.)
	WriteUTF8ResourceString	Writes a resource string to an HtmlTextWriter control. The WriteUTF8ResourceString(HtmlTextWriter,

		Int32, Int32, Boolean) method is used by generated classes and is not intended for use from within your code. (Inherited from TemplateControl.)
	XPath(String)	Evaluates an XPath data-binding expression. (Inherited from TemplateControl.)
	XPath(String, IXmlNamespaceResolver)	Evaluates an XPath data-binding expression using the specified prefix and namespace mappings for namespace resolution. (Inherited from TemplateControl.)
	XPath(String, String)	Evaluates an XPath data-binding

	expression using the specified format string to display the result. (Inherited from TemplateControl.)
XPath(String, String, IXmlNamespaceResolver)	Evaluates an XPath data-binding expression using the specified prefix and namespace mappings for namespace resolution and the specified format string to display the result. (Inherited from TemplateControl.)
XPathSelect(String)	Evaluates an XPath data-binding expression and returns a node collection that implements the IEnumerable interface. (Inherited from TemplateControl.)

XPathSelect(String, IXmlNamespaceResolver)	Evaluates an XPath data-binding expression using the specified prefix and namespace mappings for namespace resolution and returns a node collection that implements the IEnumerable interface. (Inherited from TemplateControl.)

Default. ButtonCancel_Click Method

ButtonCancel_Click.

Namespace: AsteriskManager

Assembly: AsteriskManager (in AsteriskManager.dll) Version: 1.0.0.0 (1.0.0.0)

Syntax

C#

VB

C++

F#

Copy

```
protected void ButtonCancel_Click(
	Object sender,
	EventArgs e
)
```

sender

Type: System. Object
what control.

e

Type: System. EventArgs
EventArgs from the control.

◢Remarks

Every good program needs a way to back-out We simply cancel the creation of a new user.

ButtonChangeVMPass_Click.
Namespace: AsteriskManager
Assembly: AsteriskManager (in AsteriskManager.dll) Version: 1.0.0.0 (1.0.0.0)

Default. ButtonChangeVMPass_Click Method

◢Syntax

C#

VB

C++

F#

Copy

```
protected void ButtonChangeVMPass_Click(
        Object sender,
        EventArgs e
)
```

Parameters

sender

Type: System. Object
what control.

e

Type: System. EventArgs
EventArgs from the control.

◢Remarks

This calls the function that changes the VM password.

Default. ButtonCreateNewUser_Click Method

ButtonCreateNewUser_Click.

Namespace: AsteriskManager

Assembly: AsteriskManager (in AsteriskManager.dll) Version: 1.0.0.0 (1.0.0.0)

Syntax

C#

VB

C++

F#

Copy

```
protected void ButtonCreateNewUser_Click(
        Object sender,
        EventArgs e
)
```

Parameters

sender

Type: System. Object
what control.

e

Type: System. EventArgs
EventArgs from the control.

Remarks

Here we hide controls and make others visible that are relivent We also change the text on the button and depending on the text We will call the function to create the new user.

Default. ButtonCreateVM_Click Method

ButtonCreateVM_Click.
Namespace: AsteriskManager
Assembly: AsteriskManager (in AsteriskManager.dll) Version: 1.0.0.0 (1.0.0.0)

Syntax

C#

VB

C++

F#

Copy

```
protected void ButtonCreateVM_Click(
        Object sender,
        EventArgs e
)
```

Parameters

sender

Type: System. Object
what control.

e

Type: System. EventArgs
EventArgs from the control.

Remarks

This is where we create a new VM Box for the number We call the function that performs the DB transactions.

Default. CheckBoxCurrentCustomer_CheckedChanged Method

CheckBoxCurrentCustomer_CheckedChanged.

Namespace: AsteriskManager

Assembly: AsteriskManager (in AsteriskManager.dll) Version: 1.0.0.0 (1.0.0.0)

Syntax

C#

VB

C++

F#

Copy

```
protected void CheckBoxCurrentCustomer_CheckedChanged(
	Object sender,
	EventArgs e
)
```

Parameters

sender

Type: System. Object
what control.

e

Type: System. EventArgs
EventArgs from the control.

Remarks

We let the program know that on the new user being created That the user belongs to a group that is currently a customer.

Default. CustomerId_SelectedIndexChanged Method

CustomerId_SelectedIndexChanged.

Namespace: AsteriskManager

Assembly: AsteriskManager (in AsteriskManager.dll) Version: 1.0.0.0 (1.0.0.0)

Syntax

C#

VB

C++

F#

Copy

```
protected void CustomerId_SelectedIndexChanged(
        Object sender,
        EventArgs e
)
```

Parameters

sender

Type: System. Object
what control.

e

Type: System. EventArgs
EventArgs from the control.

◢Remarks

when we select a customer ID this populates customer number(s).

Default. CustomerNumberDL_SelectedIndexChanged Method

CustomerNumberDL_SelectedIndexChanged.

Namespace: AsteriskManager

Assembly: AsteriskManager (in AsteriskManager.dll) Version: 1.0.0.0 (1.0.0.0)

Syntax

C#

VB

C++

F#

Copy

```
protected void CustomerNumberDL_SelectedIndexChanged(
	Object sender,
	EventArgs e
)
```

Parameters

sender

Type: System. Object
what control.

e

Type: System. EventArgs
EventArgs from the control.

◢Remarks

When a customer has more then one device this is visible and lets you select what number you need.

Default. CustomerNumberTXT_TextChanged Method

CustomerNumberTXT_TextChanged.

Namespace: AsteriskManager

Assembly: AsteriskManager (in AsteriskManager.dll) Version: 1.0.0.0 (1.0.0.0)

Syntax

C#

VB

C++

F#

Copy

```
protected void CustomerNumberTXT_TextChanged(
	Object sender,
	EventArgs e
)
```

Parameters

sender

Type: System. Object
what control.

e

Type: System. EventArgs
EventArgs from the control.

◢Remarks

We will add code here to ensure that we only have the type of data we want.

Default. fill_Drop_Down Method

fill_Drop_Down.

Namespace: AsteriskManager

Assembly: AsteriskManager (in AsteriskManager.dll) Version: 1.0.0.0 (1.0.0.0)

◢Syntax

C#

VB

C++

F#

Copy

```
public void fill_Drop_Down()
```

◢Remarks

This void fills the main dropdown with the default customer ID's.

Default. Page_Load Method

Page_Load.

Namespace: AsteriskManager

Assembly: AsteriskManager (in AsteriskManager.dll) Version: 1.0.0.0 (1.0.0.0)

◢Syntax

C#

VB

C++

F#

Copy

```
protected void Page_Load(
	Object sender,
	EventArgs e
)
```

Parameters

sender

Type: System. Object

what control.

e

Type: System. EventArgs

EventArgs from the control.

Remarks

This is where we load the controls and set the properties
Where needed for authentication and special rules.

Default Properties

The Default type exposes the following members.

Properties

	Name	Description
	Adapter	Gets the browser-specific adapter for the control. (Inherited from Control.)
	Application	Gets the HttpApplicationState object for the current Web request. (Inherited from Page.)

	AppRelativeTemplateSourceDirectory	Gets or sets the application-relative virtual directory of the Page or UserControl object that contains this control. (Inherited from Control.)
	AppRelativeVirtualPath	Gets or sets the application-relative, virtual directory path to the file from which the control is parsed and compiled. (Inherited from TemplateControl.)
	AspCompatMode	Sets a value indicating whether the

		page can be executed on a single-threaded apartment (STA) thread. (Inherited from Page.)
	AsyncMode	Sets a value indicating whether the page is processed synchronously or asynchronously. (Inherited from Page.)
	AsyncTimeout	Gets or sets a value indicating the time-out interval used when processing asynchronous tasks. (Inherited from

		Page.)
	AutoHandlers	**Obsolete.** The AutoHandlers property has been deprecated in ASP.NET NET 2.0. It is used by generated classes and is not intended for use within your code. (Inherited from TemplateControl.)
	AutoPostBackControl	Gets or sets the control in the page that is used to perform postbacks. (Inherited from Page.)

	BindingContainer	Gets the control that contains this control's data binding. (Inherited from Control.)
	Buffer	Sets a value indicating whether the page output is buffered. (Inherited from Page.)
	Cache	Gets the Cache object associated with the application in which the page resides. (Inherited from Page.)
	ChildControlsCreated	Gets a value that

		indicates whether the server control's child controls have been created. (Inherited from Control.)
	ClientID	Gets the control ID for HTML markup that is generated by ASP.NET. (Inherited from Control.)
	ClientIDMode	Gets or sets the algorithm that is used to generate the value of the ClientID property. (Inherited from Control.)

	ClientIDSeparator	Gets a character value representing the separator character used in the ClientID property. (Inherited from Control.)
	ClientQueryString	Gets the query string portion of the requested URL. (Inherited from Page.)
	ClientScript	Gets a ClientScriptManager object used to manage, register, and add script to the page. (Inherited from Page.)

	ClientTarget	Gets or sets a value that allows you to override automatic detection of browser capabilities and to specify how a page is rendered for particular browser clients. (Inherited from Page.)
	CodePage	Sets the code page identifier for the current Page. (Inherited from Page.)
	ContentType	Sets the HTTP MIME type for the HttpResponse object associated

		with the page. (Inherited from Page.)
	Context	Gets the HttpContext object associated with the page. (Inherited from Page.)
	Controls	Gets a ControlCollection object that represents the child controls for a specified server control in the UI hierarchy. (Inherited from Control.)
	Culture	Sets the culture ID for the Thread object associated

		with the page. (Inherited from Page.)
	DataItemContainer	Gets a reference to the naming container if the naming container implements IDataItemContainer. (Inherited from Control.)
	DataKeysContainer	Gets a reference to the naming container if the naming container implements IDataKeysControl. (Inherited from Control.)

	DesignMode	Gets a value indicating whether a control is being used on a design surface. (Inherited from Control.)
	EnableEventValidation	Gets or sets a value indicating whether the page validates postback and callback events. (Inherited from Page.)
	EnableTheming	Gets or sets a Boolean value indicating whether themes apply to the control that is derived from the TemplateControl

		class. (Inherited from TemplateControl.)
	EnableViewState	Gets or sets a value indicating whether the page maintains its view state, and the view state of any server controls it contains, when the current page request ends. (Inherited from Page.)
	EnableViewStateMac	Gets or sets a value indicating whether ASP.NET should check message authentication codes (MAC) in the page's view

		state when the page is posted back from the client. (Inherited from Page.)
	ErrorPage	Gets or sets the error page to which the requesting browser is redirected in the event of an unhandled page exception. (Inherited from Page.)
	Events	Gets a list of event handler delegates for the control. This property is read-only. (Inherited from

		Control.)
	FileDependencies	**Obsolete.** Sets an array of files that the current HttpResponse object is dependent upon. (Inherited from Page.)
	Form	Gets the HTML form for the page. (Inherited from Page.)
	HasChildViewState	Gets a value indicating whether the current server control's child controls have any saved view-

		state settings. (Inherited from Control.)
	Header	Gets the document header for the page if the head element is defined with a runat=server in the page declaration. (Inherited from Page.)
	ID	Gets or sets an identifier for a particular instance of the Page class. (Inherited from Page.)
	IdSeparator	Gets the character used to

		separate control identifiers when building a unique ID for a control on a page. (Inherited from Page.)
	IsAsync	Gets a value indicating whether the page is processed asynchronously. (Inherited from Page.)
	IsCallback	Gets a value that indicates whether the page request is the result of a callback. (Inherited from Page.)

	IsChildControlStateCleared	Gets a value indicating whether controls contained within this control have control state. (Inherited from Control.)
	IsCrossPagePostBack	Gets a value indicating whether the page is involved in a cross-page postback. (Inherited from Page.)
	IsPostBack	Gets a value that indicates whether the page is being rendered for the first time or is being loaded in response to a

		postback. (Inherited from Page.)
	IsPostBackEventControlRegistered	Gets a value that indicates whether the control in the page that performs postbacks has been registered. (Inherited from Page.)
	IsReusable	Gets a value indicating whether the Page object can be reused. (Inherited from Page.)
	IsTrackingViewState	Gets a value that indicates whether the

		server control is saving changes to its view state. (Inherited from Control.)
	IsValid	Gets a value indicating whether page validation succeeded. (Inherited from Page.)
	IsViewStateEnabled	Gets a value indicating whether view state is enabled for this control. (Inherited from Control.)
	Items	Gets a list of objects stored in the page

		context. (Inherited from Page.)
	LCID	Sets the locale identifier for the Thread object associated with the page. (Inherited from Page.)
	LoadViewStateByID	Gets a value indicating whether the control participates in loading its view state by ID instead of index. (Inherited from Control.)
	MaintainScrollPositionOnPostBack	Gets or sets a value indicating whether to

		return the user to the same position in the client browser after postback. This property replaces the obsolete SmartNavigation property. (Inherited from Page.)
	Master	Gets the master page that determines the overall look of the page. (Inherited from Page.)
	MasterPageFile	Gets or sets the virtual path of the master page. (Inherited from Page.)

	MaxPageStateFieldLength	Gets or sets the maximum length for the page's state field. (Inherited from Page.)
	MetaDescription	Gets or sets the content of the "description" meta element. (Inherited from Page.)
	MetaKeywords	Gets or sets the content of the "keywords" meta element. (Inherited from Page.)
	NamingContainer	Gets a reference to the server control's naming container, which

		creates a unique namespace for differentiating between server controls with the same ID property value. (Inherited from Control.)
	Page	Gets a reference to the Page instance that contains the server control. (Inherited from Control.)
	PageAdapter	Gets the adapter that renders the page for the specific requesting browser. (Inherited from Page.)

	PageStatePersister	Gets the PageStatePersister object associated with the page. (Inherited from Page.)
	Parent	Gets a reference to the server control's parent control in the page control hierarchy. (Inherited from Control.)
	PreviousPage	Gets the page that transferred control to the current page. (Inherited from Page.)
	RenderingCompatibility	Gets a value that

		specifies the ASP.NET version that rendered HTML will be compatible with. (Inherited from Control.)
	Request	Gets the HttpRequest object for the requested page. (Inherited from Page.)
	Response	Gets the HttpResponse object associated with the Page object. This object allows you to send HTTP response data to a client and contains information about that

		response. (Inherited from Page.)
	ResponseEncoding	Sets the encoding language for the current HttpResponse object. (Inherited from Page.)
	RouteData	Gets the RouteData value of the current RequestContext instance. (Inherited from Page.)
	Server	Gets the Server object, which is an instance of the HttpServerUtility

		class. (Inherited from Page.)
	Session	Gets the current Session object provided by ASP.NET. (Inherited from Page.)
	Site	Gets information about the container that hosts the current control when rendered on a design surface. (Inherited from Control.)
	SkinID	Gets or sets the skin to apply to the control. (Inherited from

		Control.)
	SmartNavigation	**Obsolete.** Gets or sets a value indicating whether smart navigation is enabled. This property is deprecated. (Inherited from Page.)
	StyleSheetTheme	Gets or sets the name of the theme that is applied to the page early in the page life cycle. (Inherited from Page.)
	SupportAutoEvents	Gets a value indicating whether the TemplateControl

		control supports automatic events. (Inherited from TemplateControl.)
	TemplateControl	Gets or sets a reference to the template that contains this control. (Inherited from Control.)
	TemplateSourceDirectory	Gets the virtual directory of the Page or UserControl that contains the current server control. (Inherited from Control.)

	Theme	Gets or sets the name of the page theme. (Inherited from Page.)
	Title	Gets or sets the title for the page. (Inherited from Page.)
	Trace	Gets the TraceContext object for the current Web request. (Inherited from Page.)
	TraceEnabled	Sets a value indicating whether tracing is enabled for the Page object. (Inherited from

		Page.)
	TraceModeValue	Sets the mode in which trace statements are displayed on the page. (Inherited from Page.)
	TransactionMode	Sets the level of transaction support for the page. (Inherited from Page.)
	UICulture	Sets the user interface (UI) ID for the Thread object associated with the page. (Inherited from Page.)

	UniqueFilePathSuffix	Gets a unique suffix to append to the file path for caching browsers. (Inherited from Page.)
	UniqueID	Gets the unique, hierarchically qualified identifier for the server control. (Inherited from Control.)
	User	Gets information about the user making the page request. (Inherited from Page.)
	Validators	Gets a collection of all validation

		controls contained on the requested page. (Inherited from Page.)
	ViewState	Gets a dictionary of state information that allows you to save and restore the view state of a server control across multiple requests for the same page. (Inherited from Control.)
	ViewStateEncryptionMode	Gets or sets the encryption mode of the view state. (Inherited from Page.)

	ViewStateIgnoresCase	Gets a value that indicates whether the StateBag object is case-insensitive. (Inherited from Control.)
	ViewStateMode	Gets or sets the view-state mode of this control. (Inherited from Control.)
	ViewStateUserKey	Assigns an identifier to an individual user in the view-state variable associated with the current page. (Inherited from Page.)

	Visible	Gets or sets a value indicating whether the Page object is rendered. (Inherited from Page.)

Default Events

The Default type exposes the following members.

Events

	Name	Description
	AbortTransaction	Occurs when a user ends a transaction. (Inherited from TemplateControl.)
	CommitTransaction	Occurs when a transaction completes. (Inherited from TemplateControl.)
	DataBinding	Occurs when the server control binds to a data source. (Inherited from

		Control.)
⚡	Disposed	Occurs when a server control is released from memory, which is the last stage of the server control lifecycle when an ASP.NET page is requested. (Inherited from Control.)
⚡	Error	Occurs when an unhandled exception is thrown. (Inherited from TemplateControl.)
⚡	Init	Occurs when the server control is initialized, which is the first step in its lifecycle. (Inherited from

		Control.)
⚡	InitComplete	Occurs when page initialization is complete. (Inherited from Page.)
⚡	Load	Occurs when the server control is loaded into the Page object. (Inherited from Control.)
⚡	LoadComplete	Occurs at the end of the load stage of the page's life cycle. (Inherited from Page.)
⚡	PreInit	Occurs at the beginning of page initialization. (Inherited from

		Page.)
	PreLoad	Occurs before the page Load event. (Inherited from Page.)
	PreRender	Occurs after the Control object is loaded but prior to rendering. (Inherited from Control.)
	PreRenderComplete	Occurs before the page content is rendered. (Inherited from Page.)
	SaveStateComplete	Occurs after the page has completed saving all view state and control state information for the

		page and controls on the page. (Inherited from Page.)
	Unload	Occurs when the server control is unloaded from memory. (Inherited from Control.)

AsteriskManager.Classes Namespace

namespace AsteriskManagert.Classes

◢Remarks

/// We use classes so that we dont have to re-write the same code over and over ///

◢Classes

	Class	Description
	Database	Database

Database Class

Database

◢Inheritance Hierarchy

System. Object
  AsteriskManager.Classes. Database

Namespace: AsteriskManager.Classes
Assembly: AsteriskManager (in AsteriskManager.dll) Version: 1.0.0.0 (1.0.0.0)

◢Syntax

C#

VB

C++

F#

Copy

```
public class Database
```

The Database type exposes the following members.

◢Constructors

	Name	Description

	Name	Description
	Database	Initializes a new instance of the Database class

Top

Methods

	Name	Description
S	ChangeCustomerVMPass	ChangeCustomerVMPass.
S	CreateCustomer	CreateCustomer.
S	CreateCustomerVM	CreateCustomerVM.
	Equals	Determines whether the specified Object is equal to the current Object. (Inherited from Object.)
S	ErrorLogmeNow	ErrorLogmeNow.

	Finalize	Allows an object to try to free resources and perform other cleanup operations before it is reclaimed by garbage collection. (Inherited from Object.)
S	GetAllCustIfo	GetAllCustIfo.
S	GetAllNums	GetAllNums.
S	GetCustomerInfo	GetCustomerInfo.
S	GetCustomerNumbers	GetCustomerNumbers.
S	GetCustomerVM	GetCustomerVM.
	GetHashCode	Serves as a hash function for a particular type.

		(Inherited from Object.)
	GetType	Gets the Type of the current instance. (Inherited from Object.)
S	LogmeNow	LogmeNow.
	MemberwiseClone	Creates a shallow copy of the current Object. (Inherited from Object.)
	ToString	Returns a string that represents the current object. (Inherited from Object.)

Top

Remarks

This is where we make calls to the DB in setting up this class file we can re-use code instead of writing the same thing over and over.

Database Constructor

Initializes a new instance of the Database class

Namespace: AsteriskManager.Classes

Assembly: AsteriskManager (in AsteriskManager.dll) Version: 1.0.0.0 (1.0.0.0)

Syntax

C#

VB

C++

F#

Copy

```
public Database()
```

Database Methods

The Database type exposes the following members.

Methods

	Name	Description
S	ChangeCustomerVMPass	ChangeCustomerVMPass.
S	CreateCustomer	CreateCustomer.
S	CreateCustomerVM	CreateCustomerVM.
	Equals	Determines whether the specified Object is equal to the current Object. (Inherited from Object.)

S	ErrorLogmeNow	ErrorLogmeNow.
	Finalize	Allows an object to try to free resources and perform other cleanup operations before it is reclaimed by garbage collection. (Inherited from Object.)
S	GetAllCustIfo	GetAllCustIfo.
S	GetAllNums	GetAllNums.
S	GetCustomerInfo	GetCustomerInfo.
S	GetCustomerNumbers	GetCustomerNumbers.
S	GetCustomerVM	GetCustomerVM.

	GetHashCode	Serves as a hash function for a particular type. (Inherited from Object.)
	GetType	Gets the Type of the current instance. (Inherited from Object.)
S	LogmeNow	LogmeNow.
	MemberwiseClone	Creates a shallow copy of the current Object. (Inherited from Object.)
	ToString	Returns a string that represents the current object. (Inherited from Object.)

Database. ChangeCustomerVMPass Method

ChangeCustomerVMPass.

Namespace: AsteriskManager.Classes

Assembly: AsteriskManager (in AsteriskManager.dll) Version: 1.0.0.0 (1.0.0.0)

Syntax

C#

VB

C++

F#

Copy

```
public static void ChangeCustomerVMPass(
        string CustomerMB,
        string NewPass
)
```

CustomerMB

Type: System. String

The Customer's New mail box number.

NewPass

Type: System. String

The Customer' new password.

◢Remarks

Changes a customers voicemail password.

Database. CreateCustomer Method

CreateCustomer.
Namespace: AsteriskManager.Classes
Assembly: AsteriskManager (in AsteriskManager.dll) Version: 1.0.0.0 (1.0.0.0)

Syntax

C#

VB

C++

F#

Copy

```
public static void CreateCustomer(
        string CustomerAccount,
        string CustomerNum,
        string Description
)
```

CustomerAccount

Type: System. String
The Customer's ID usually the main phone number.

CustomerNum

Type: System. String

The Customer's phone number.

Description

Type: System. String

The Customer's name.

Remarks

Creates a new customer and or a new customer phone number.

Database. CreateCustomerVM Method

CreateCustomerVM.

Namespace: AsteriskManager.Classes

Assembly: AsteriskManager (in AsteriskManager.dll) Version: 1.0.0.0 (1.0.0.0)

◢Syntax

C#

VB

C++

F#

Copy

```
public static void CreateCustomerVM(
        string CustomerAccount,
        string CustomerNum,
        string MBNum,
        string MBPass
)
```

CustomerAccount

Type: System. String

The Customer's ID usually the main phone number.

CustomerNum

Type: System. String

The Customer's phone number.

MBNum

Type: System. String

The Customer's mail box number.

MBPass

Type: System. String

The Customer's mail box password.

◢Remarks

Create a new voicemail box for a customer.

Database. ErrorLogmeNow Method

ErrorLogmeNow.

Namespace: AsteriskManager.Classes

Assembly: AsteriskManager (in AsteriskManager.dll) Version: 1.0.0.0 (1.0.0.0)

Syntax

C#

VB

C++

F#

Copy

```
public static void ErrorLogmeNow(
        string EFunction,
        string EError,
        string EUser,
        string PageName
)
```

EFunction

Type: System. String

Current Function that was called.

EError

Type: System. String

What was happening.

EUser

Type: System. String

The current user.

PageName

Type: System. String

What page is the user on.

◢Remarks

We can log errors here Again End-Users lie.

Database. GetAllCustIfo Method

GetAllCustIfo.

Namespace: AsteriskManager.Classes

Assembly: AsteriskManager (in AsteriskManager.dll) Version: 1.0.0.0 (1.0.0.0)

◢Syntax

C#

VB

C++

F#

Copy

```
public static DataSet GetAllCustIfo(
        string CustomerId
)
```

CustomerId

Type: System. String
The Customer ID, generally the main phone number for the customer.

Type: DataSet
The DataSet that we will bind too.

◢Remarks

Gets all the information of a customer's ID.

Database. GetAllNums Method

GetAllNums.

Namespace: AsteriskManager.Classes

Assembly: AsteriskManager (in AsteriskManager.dll) Version: 1.0.0.0 (1.0.0.0)

Syntax

C#

VB

C++

F#

Copy

```
public static DataSet GetAllNums()
```

Type: DataSet
The DataSet that we will bind too.

Remarks

Simple way to get all of the customer Id's.

Database. GetCustomerInfo Method

GetCustomerInfo.

Namespace: AsteriskManager.Classes

Assembly: AsteriskManager (in AsteriskManager.dll) Version: 1.0.0.0 (1.0.0.0)

Syntax

C#

VB

C++

F#

Copy

```
public static DataSet GetCustomerInfo(
        string CustomerId
)
```

CustomerId

Type: System. String

The Customer ID, generally the main phone number for

the customer.

Type: DataSet

The DataSet that we will bind too.

◢Remarks

Where we get the information from the customer.

Database. GetCustomerNumbers Method

GetCustomerNumbers.

Namespace: AsteriskManager.Classes

Assembly: AsteriskManager (in AsteriskManager.dll) Version: 1.0.0.0 (1.0.0.0)

◢Syntax

C#

VB

C++

F#

Copy

```
public static DataSet GetCustomerNumbers(
	string CustomerId
)
```

CustomerId

Type: System. String

The Customer ID, generally the main phone number for

the customer.

Type: DataSet
The DataSet that we will bind too.

◢Remarks

Gets all of the customer numbers associated with a customer's ID.

Database. GetCustomerVM Method

GetCustomerVM.

Namespace: AsteriskManager.Classes

Assembly: AsteriskManager (in AsteriskManager.dll) Version: 1.0.0.0 (1.0.0.0)

Syntax

C#

VB

C++

F#

Copy

```
public static DataSet GetCustomerVM(
	string CustomerId
)
```

CustomerId

Type: System. String

The Customer ID, generally the main phone number for the customer.

Type: DataSet

The DataSet that we will bind too.

◢Remarks

Gets the voice mail for a customer's phone number.

Database. LogmeNow Method

LogmeNow.

Namespace: AsteriskManager.Classes

Assembly: AsteriskManager (in AsteriskManager.dll) Version: 1.0.0.0 (1.0.0.0)

Syntax

C#

VB

C++

F#

```
public static void LogmeNow(
	string UserName,
	string actions,
	string Variables,
	string PageId
)
```

UserName

Type: System. String
Current user.

actions

Type: System. String

What was happening.

Variables

Type: System. String

The issue at hand.

PageId

Type: System. String

What page is the user on.

◢Remarks

Where we can send logs to the database Because end-users lie.

Source Code

In the following pages you will see the source code that is used to create a simple web interface for managing the Asterisk PBX system that is setup the same way as it was in the previous book.

This is just a guide for you but by all means feel free to change it and add on to it as there are limitless possibilities for you to grow this into an enterprise application.

Default.aspx

```
<%@ Page Language="C#" AutoEventWireup="true"
CodeBehind="Default.aspx.cs"
Inherits="AsteriskManager.Default" %>

<!DOCTYPE html>

<html xmlns="http://www.w3.org/1999/xhtml">
<head runat="server">
    <title></title>
    <style>

    .Section1 {
    line-height:30px;
    height:100px;
    width:100px;
    float:left;
    padding:0px;
    }
    .Section2 {
    line-height:30px;
    height:428px;
    width:795px;
    position:center;
     margin: 0px auto;
    padding:10px;
    }
     .Section3 {
    line-height:30px;
    height:100px;
    width:110px;
    float:left;
    padding:0px;
    }
      .Section4 {
    line-height:30px;
    height:100px;
    width:300px;
```

```
    float:left;
    padding:5px;
    }
        </style>


</head>
<body>
    <form id="form1" runat="server">
    <div>

        <asp:ScriptManager ID="ScriptManager1" runat="server">
        </asp:ScriptManager>
        <asp:UpdatePanel ID="UpdatePanel1" runat="server">
            <ContentTemplate>
                <br />
                <br />
                <br />
                <div class="Section2">
                <div class="Section1">
                     Customer ID
                    <br />
                <asp:DropDownList ID="CustomerId"
runat="server" AutoPostBack="True"
OnSelectedIndexChanged="CustomerId_SelectedIndexChanged"
Width="100px">
                </asp:DropDownList>
                <asp:TextBox ID="TextBoxNCustID"
runat="server" Width="85px" Visible="False">Customer
ID</asp:TextBox>

                </div>
                <div class="Section1">
                <asp:Label ID="LabelPN" runat="server"
Text="Phone Number"></asp:Label>
                     <br />
                <asp:TextBox ID="TextBoxNewNumber"
runat="server" Width="85px"
Visible="False">Number</asp:TextBox>

                <asp:TextBox ID="CustomerNumberTXT"
runat="server" Enabled="False"
OnTextChanged="CustomerNumberTXT_TextChanged" MaxLength="10"
Width="85px"></asp:TextBox>

                <asp:DropDownList ID="CustomerNumberDL"
runat="server" AutoPostBack="True" Width="100px"
OnSelectedIndexChanged="CustomerNumberDL_SelectedIndexChanged"
```

```
>
                <asp:DropDownList>
                </div>
                <div class="Section3">
                <asp:Label ID="LabelVM" runat="server"
Text="Voice Mail Box"></asp:Label>
                <br />
                <asp:TextBox ID="VoiceMailNumber"
runat="server" Width="85px" Enabled="False"></asp:TextBox>
                </div>
                <div class="Section1">
                <asp:Label ID="LabelVMPass" runat="server"
Text="VM Password"></asp:Label>
                <br />
                <asp:TextBox ID="VoiceMailPassTx"
runat="server" Width="85px"></asp:TextBox>
                </div>

                <br />
                    <br /><br /><br />
                <div class="Section4">
                    <asp:Label ID="LabelName" runat="server"
Text="Name"></asp:Label>
                    <br />
                    <asp:TextBox ID="TextboxName"
runat="server" Width="223px" Enabled="False"></asp:TextBox>
                </div>
                    <br />
                    <br />
                    <br />
                    <br />
                    <asp:CheckBox ID="CheckBoxCurrentCustomer"
runat="server" Text="Current Customer" AutoPostBack="True"
OnCheckedChanged="CheckBoxCurrentCustomer_CheckedChanged"
Visible="False" />
                    <br />
                    <br />
                    <asp:Button ID="ButtonCreateVM"
runat="server" Text="Create VM Account" Width="340px"
OnClick="ButtonCreateVM_Click" />
                    <br />
                    <asp:Button ID="ButtonChangeVMPass"
runat="server" OnClick="ButtonChangeVMPass_Click" Text="Change
VM Password" Width="340px" />
                    <br />
                    <asp:Button ID="ButtonCreateNewUser"
runat="server" OnClick="ButtonCreateNewUser_Click"
Text="Create New User" Width="340px" />
```

```
                     <asp:Button ID="ButtonCancel"
runat="server" OnClick="ButtonCancel_Click" Text="Cancel"
Visible="False" />
                </div>
                <br />
                <br />
                <br />
                <br />
                <br />
                <br />
                <br />
                <br />
                <br />
                <br />
                <br />
                <br />
                <br />
                <br />
                <br />
                <br />
                <br />
            </ContentTemplate>
        </asp:UpdatePanel>

    </div>
    </form>
</body>
</html>
```

Default.aspx.cs

```
using System;
using AsteriskManager.Classes;
namespace AsteriskManager
{
    /// <summary>
    /// Partial class Default.
    /// </summary>
    /// <remarks>
    /// Simply half of the full class
    /// one half being the web page and the other being the
code behind
    /// </remarks>

    public partial class Default : System.Web.UI.Page
    {
        /// <summary>
        /// Page_Load.
        /// </summary>
        /// <param name="sender">what control.</param>object
        /// <param name="e">EventArgs from the
control.</param>EventArgs
        /// <remarks>
        /// This is where we load the controls and set the
properies
        /// Where needed for authentication and special rules
        /// </remarks>

        protected void Page_Load(object sender, EventArgs e)
        {
            if (!IsPostBack)
            {
                CustomerNumberDL.Visible = false;
                CustomerNumberTXT.Visible = false;
                VoiceMailNumber.Visible = false;
                VoiceMailPassTx.Visible = false;
                LabelPN.Visible = false;
                LabelVM.Visible = false;
```

```
                LabelVMPass.Visible = false;
                fill_Drop_Down();
            }
            else
            {

            }
        }
        /// <summary>
        /// fill_Drop_Down.
        /// </summary>
        /// <param name="sender">what control.</param>object
        /// <param name="e">EventArgs from the
control.</param>EventArgs
        /// <remarks>
        /// This void fills the main dropdown
        /// with the default customer ID's
        /// </remarks>

        public void fill_Drop_Down()
        {
            try
            {
                CustomerId.Dispose();
                CustomerId.DataTextField = "accountcode";
                CustomerId.DataValueField = "accountcode";
                CustomerId.DataSource = Database.GetAllNums();
                CustomerId.DataBind();
                string CustNum =
CustomerId.SelectedValue.ToString();
                int DSCount =
Database.GetCustomerNumbers(CustNum).Tables[0].Rows.Count;
                if (DSCount > 0)
                {
                    if (DSCount == 1)
                    {
                        string TmpText =
Database.GetAllCustIfo(CustNum).Tables[0].Rows[0].ItemArray.Ge
tValue(1).ToString();
                        CustomerNumberTXT.Text = TmpText;
                        CustomerNumberTXT.Visible = true;
                        LabelPN.Visible = true;
                        string vmbox =
Database.GetAllCustIfo(CustNum).Tables[0].Rows[0].ItemArray.Ge
tValue(6).ToString();
                        string TmpNMText =
Database.GetAllCustIfo(CustNum).Tables[0].Rows[0].ItemArray.Ge
tValue(42).ToString();
```

```
                        TextboxName.Text = TmpNMText;

                        if (vmbox != "")
                        {
                            string TmpVMText =
Database.GetCustomerVM(CustNum).Tables[0].Rows[0].ItemArray.Ge
tValue(0).ToString();
                            string TmpVMTpassext =
Database.GetCustomerVM(TmpVMText).Tables[0].Rows[0].ItemArray.
GetValue(1).ToString();
                            LabelVM.Visible = true;
                            LabelVMPass.Visible = true;
                            VoiceMailPassTx.Text =
TmpVMTpassext;
                            VoiceMailNumber.Text = TmpVMText;
                            VoiceMailNumber.Visible = true;
                            VoiceMailPassTx.Visible = true;
                            VoiceMailNumber.Enabled = false;
                            ButtonCreateVM.Enabled = false;
                            ButtonChangeVMPass.Enabled = true;

                        }
                        else
                        {
                            LabelVM.Visible = false;
                            LabelVMPass.Visible = false;
                            VoiceMailNumber.Visible = false;
                            VoiceMailPassTx.Visible = false;
                            ButtonCreateVM.Enabled = true;
                            ButtonChangeVMPass.Enabled =
false;


                        }

                    }
                    else
                    {
                        if (DSCount > 1)
                        {
                            CustomerNumberTXT.Visible = false;
                        }
                    }
                }
            }
            catch (Exception ex)
            {
                string EX = ex.Message;
```

```
            }
        }
        /// <summary>
        /// CustomerId_SelectedIndexChanged.
        /// </summary>
        /// <param name="sender">what control.</param>object
        /// <param name="e">EventArgs from the
control.</param>EventArgs
        /// <remarks>
        /// when we select a customer ID
        /// this populates customer number(s)
        /// </remarks>

        protected void CustomerId_SelectedIndexChanged(object
sender, EventArgs e)
        {
            if (CheckBoxCurrentCustomer.Checked == true)
            {
                return;
            }
            else
            {
                string CustNum =
CustomerId.SelectedValue.ToString();

                int DSCount =
Database.GetCustomerNumbers(CustNum).Tables[0].Rows.Count;
                if (DSCount > 0)
                {
                    if (DSCount != 1)
                    {
                        CustomerNumberDL.Dispose();
                        CustomerNumberDL.DataTextField =
"Name";
                        CustomerNumberDL.DataValueField =
"Name";
                        CustomerNumberDL.DataSource =
Database.GetCustomerNumbers(CustNum);
                        CustomerNumberDL.DataBind();
                        CustomerNumberDL.Visible = true;
                        CustomerNumberTXT.Visible = false;
                        string CustNumDD =
CustomerNumberDL.SelectedValue.ToString();
                        string TmpNMText =
Database.GetAllCustIfo(CustNumDD).Tables[0].Rows[0].ItemArray.
GetValue(42).ToString();
                        TextboxName.Text = TmpNMText;
```

```
                        string vmbox =
Database.GetAllCustIfo(CustNumDD).Tables[0].Rows[0].ItemArray.
GetValue(6).ToString();
                        if (vmbox != "")
                        {
                            string TmpVMText =
Database.GetCustomerVM(CustNumDD).Tables[0].Rows[0].ItemArray.
GetValue(0).ToString();
                            string TmpVMTpassext =
Database.GetCustomerVM(CustNumDD).Tables[0].Rows[0].ItemArray.
GetValue(1).ToString();
                            LabelVM.Visible = true;
                            LabelVMPass.Visible = true;
                            VoiceMailPassTx.Text =
TmpVMTpassext;
                            VoiceMailNumber.Text = TmpVMText;
                            VoiceMailNumber.Visible = true;
                            VoiceMailPassTx.Visible = true;
                            VoiceMailNumber.Enabled = false;
                            ButtonCreateVM.Enabled = false;
                            ButtonChangeVMPass.Enabled = true;

                        }
                        else
                        {
                            LabelVM.Visible = false;
                            VoiceMailNumber.Visible = false;
                            VoiceMailPassTx.Visible = false;
                            ButtonCreateVM.Enabled = true;
                            ButtonChangeVMPass.Enabled =
false;


                        }

                    }
                    else
                    {
                        string TmpNMText =
Database.GetAllCustIfo(CustNum).Tables[0].Rows[0].ItemArray.Ge
tValue(42).ToString();
                        TextboxName.Text = TmpNMText;

                        string TmpText =
Database.GetAllCustIfo(CustNum).Tables[0].Rows[0].ItemArray.Ge
tValue(1).ToString();
```

```
                        CustomerNumberTXT.Text = TmpText;
                        CustomerNumberTXT.Visible = true;
                        LabelPN.Visible = true;
                        CustomerNumberDL.Visible = false;
                        string vmbox =
Database.GetAllCustIfo(CustNum).Tables[0].Rows[0].ItemArray.Ge
tValue(6).ToString();
                        if (vmbox != "")
                        {
                            string TmpVMText =
Database.GetCustomerVM(CustNum).Tables[0].Rows[0].ItemArray.Ge
tValue(0).ToString();
                            string TmpVMTpassext =
Database.GetCustomerVM(TmpVMText).Tables[0].Rows[0].ItemArray.
GetValue(1).ToString();
                            LabelVM.Visible = true;
                            LabelVMPass.Visible = true;
                            VoiceMailPassTx.Text =
TmpVMTpassext;
                            VoiceMailNumber.Text = TmpVMText;
                            VoiceMailNumber.Visible = true;
                            VoiceMailPassTx.Visible = true;
                            VoiceMailNumber.Enabled = false;
                            ButtonCreateVM.Enabled = false;
                            ButtonChangeVMPass.Enabled = true;

                        }
                        else
                        {
                            ButtonCreateVM.Enabled = true;
                            ButtonChangeVMPass.Enabled =
false;


                            LabelVM.Visible = false;
                            LabelVMPass.Visible = false;
                            VoiceMailNumber.Visible = false;
                            VoiceMailPassTx.Visible = false;

                        }
                    }

                }
                else
                {
                    CustomerNumberDL.Visible = false;
                    CustomerNumberTXT.Visible = false;
                    LabelPN.Visible = true;
                }
```

```
            }
        }
        /// <summary>
        /// CustomerNumberTXT_TextChanged.
        /// </summary>
        /// <param name="sender">what control.</param>object
        /// <param name="e">EventArgs from the
control.</param>EventArgs
        /// <remarks>
        /// We will add code here to ensure that we only have
the type of data
        /// we want.
        /// </remarks>

        protected void CustomerNumberTXT_TextChanged(object
sender, EventArgs e)
        {

        }
        /// <summary>
        /// CustomerNumberDL_SelectedIndexChanged.
        /// </summary>
        /// <param name="sender">what control.</param>object
        /// <param name="e">EventArgs from the
control.</param>EventArgs
        /// <remarks>
        /// When a customer has more then one device
        /// this is vissible and lets you select what number
you need
        /// </remarks>

        protected void
CustomerNumberDL_SelectedIndexChanged(object sender, EventArgs
e)
        {
            string CustNumDD =
CustomerNumberDL.SelectedValue.ToString();
            string vmbox =
Database.GetAllCustIfo(CustNumDD).Tables[0].Rows[0].ItemArray.
GetValue(6).ToString();
            string TmpNMText =
Database.GetAllCustIfo(CustNumDD).Tables[0].Rows[0].ItemArray.
GetValue(42).ToString();
            TextboxName.Text = TmpNMText;

            if (vmbox != "")
            {
                string TmpVMText =
```

```
Database.GetCustomerVM(CustNumDD).Tables[0].Rows[0].ItemArray.
GetValue(0).ToString();
                    string TmpVMTpassext =
Database.GetCustomerVM(CustNumDD).Tables[0].Rows[0].ItemArray.
GetValue(1).ToString();
                    LabelVM.Visible = true;
                    LabelVMPass.Visible = true;
                    VoiceMailPassTx.Text = TmpVMTpassext;
                    VoiceMailNumber.Text = TmpVMText;
                    VoiceMailNumber.Visible = true;
                    VoiceMailPassTx.Visible = true;
                    VoiceMailNumber.Enabled = false;
                    ButtonCreateVM.Enabled = false;
                    ButtonChangeVMPass.Enabled = true;

                }
                else
                {
                    ButtonCreateVM.Enabled = true;
                    ButtonChangeVMPass.Enabled = false;
                    LabelVM.Visible = false;
                    LabelVMPass.Visible = false;
                    VoiceMailNumber.Visible = false;
                    VoiceMailPassTx.Visible = false;

                }
            }
            /// <summary>
            /// ButtonChangeVMPass_Click.
            /// </summary>
            /// <param name="sender">what control.</param>object
            /// <param name="e">EventArgs from the
control.</param>EventArgs
            /// <remarks>
            /// This calls the function that changes the VM
password
            /// </remarks>

            protected void ButtonChangeVMPass_Click(object sender,
EventArgs e)
            {

Database.ChangeCustomerVMPass(VoiceMailNumber.Text,
VoiceMailPassTx.Text);
            }
            /// <summary>
            /// ButtonCreateNewUser_Click.
            /// </summary>
```

```
        /// <param name="sender">what control.</param>object
        /// <param name="e">EventArgs from the
control.</param>EventArgs
        /// <remarks>
        /// Here we hide controls and make others visible that
are relivent
        /// We also change the text on the button and
depending on the text
        /// We will call the function to create the new user
        /// </remarks>

        protected void ButtonCreateNewUser_Click(object
sender, EventArgs e)
        {
            if (ButtonCreateNewUser.Text == "Create New User")
            {
                ButtonCancel.Visible = true;
                ButtonChangeVMPass.Enabled = false;
                ButtonCreateVM.Enabled = false;
                VoiceMailNumber.Visible = false;
                VoiceMailPassTx.Visible = false;
                LabelVM.Visible = false;
                LabelVMPass.Visible = false;
                TextBoxNewNumber.Visible = true;
                TextboxName.Enabled = true;
                TextboxName.Text = "";
                ButtonCreateNewUser.Text = "Submit";
                CustomerNumberTXT.Visible = false;
                CheckBoxCurrentCustomer.Visible = true;
                CustomerNumberDL.Visible = false;
                if (CheckBoxCurrentCustomer.Checked == true)
                {
                    CustomerId.Visible = true;
                    TextBoxNCustID.Visible = false;
                }
                else
                {
                    CustomerId.Visible = false;
                    TextBoxNCustID.Visible = true;
                    CheckBoxCurrentCustomer.Visible = true;
                }
            }
            else
            {
                string ACCNum = "";
                String NewNum = TextBoxNewNumber.Text;
                string CustDescription = TextboxName.Text;
```

```
                if(CheckBoxCurrentCustomer.Checked==true)
                {
                    ACCNum = CustomerId.Text;
                    Database.CreateCustomer(ACCNum, NewNum,
CustDescription);
                }
                else
                {
                    ACCNum = TextBoxNCustID.Text;
                    Database.CreateCustomer(ACCNum, NewNum,
CustDescription);

                }
                ButtonCancel.Visible = false;
                ButtonCreateNewUser.Text = "Create New User";
                CheckBoxCurrentCustomer.Checked = false;
                CheckBoxCurrentCustomer.Visible = false;
                TextBoxNewNumber.Visible = false;
                TextBoxNCustID.Visible = false;
                TextboxName.Enabled = false;
                TextboxName.Text = "";
                CustomerId.Visible = true;
                CustomerNumberTXT.Visible = true;
                fill_Drop_Down();

            }
        }
        /// <summary>
        /// ButtonCreateVM_Click.
        /// </summary>
        /// <param name="sender">what control.</param>object
        /// <param name="e">EventArgs from the
control.</param>EventArgs
        /// <remarks>
        /// This is where we create a new VM Box for the
number
        /// We call the function that performs the DB
transactions
        /// </remarks>

        protected void ButtonCreateVM_Click(object sender,
EventArgs e)
        {

                if (ButtonCreateVM.Text == "Create VM
Account")
                {
                VoiceMailNumber.Text = "";
```

```
                VoiceMailPassTx.Text = "";
                VoiceMailNumber.Visible = true;
                VoiceMailNumber.Enabled = true;
                VoiceMailPassTx.Visible = true;
                ButtonCreateVM.Text = "Click to Submit";
                }
                else
                {
                string cnum = "";
                if(CustomerNumberDL.Visible== true)
                {
                     cnum =
CustomerNumberDL.SelectedValue.ToString();
                }
                else
                {
                     cnum = CustomerNumberTXT.Text;
                }
                if (cnum != "")
                {
                    string cact =
CustomerId.SelectedValue.ToString();
                    string BoxNum = VoiceMailNumber.Text;
                    string VMpass = VoiceMailPassTx.Text;
                    Database.CreateCustomerVM(cact, cnum,
BoxNum, VMpass);
                }
                VoiceMailNumber.Enabled = false;
                ButtonCreateVM.Text = "Create VM Account";
                ButtonChangeVMPass.Enabled = true;
                ButtonCreateVM.Enabled = false;
                }

        }
        /// <summary>
        /// CheckBoxCurrentCustomer_CheckedChanged.
        /// </summary>
        /// <param name="sender">what control.</param>object
        /// <param name="e">EventArgs from the
control.</param>EventArgs
        /// <remarks>
        /// We let the program know that on the new user being
created
        /// That the user belongs to a group that is currently
a customer
        /// </remarks>

        protected void
```

```
CheckBoxCurrentCustomer_CheckedChanged(object sender,
EventArgs e)
        {
            if(CheckBoxCurrentCustomer.Checked==true)
            {
                CustomerId.Visible = true;
                TextBoxNCustID.Visible = false;
            }
            else
            {
                CustomerId.Visible = false;
                TextBoxNCustID.Visible = true;
            }
        }
        /// <summary>
        /// ButtonCancel_Click.
        /// </summary>
        /// <param name="sender">what control.</param>object
        /// <param name="e">EventArgs from the
control.</param>EventArgs
        /// <remarks>
        /// Every good program needs a way to backout
        /// We simply cancel the creation of a new user
        /// </remarks>

        protected void ButtonCancel_Click(object sender,
EventArgs e)
        {
            ButtonCancel.Visible = false;
            ButtonCreateNewUser.Text = "Create New User";
            CheckBoxCurrentCustomer.Checked = false;
            CheckBoxCurrentCustomer.Visible = false;
            TextBoxNewNumber.Visible = false;
            TextBoxNCustID.Visible = false;
            TextboxName.Enabled = false;
            TextboxName.Text = "";
            CustomerId.Visible = true;
            CustomerNumberTXT.Visible = true;
            fill_Drop_Down();
        }
    }
}
```

Database.cs

```
using System;
using System.Collections.Generic;
using MySql.Data.MySqlClient;
using System.Data;
namespace AsteriskManager.Classes
{
    /// <summary>
    /// Database
    /// </summary>
    /// <remarks>
    /// This is where we make calls to the DB
    /// In setting up this class file we can re-use code
instead of writing the samething
    /// over and over
    /// </remarks>

    public class Database
    {
        //Logging
        static private string LoggingManagerHost = "The IP
address of your log server";
        static private string LoggingManagerDatabase =
"ApplicationLogs";
        static private string LoggingManagerUser = "MySqlo
Usert";
        static private string LoggingManagerPassword = "MySql
pass";
        ///This is for the App Database
        static private string ApplicationManagerHost = " The
IP address of your Asterisk server ";
        static private string ApplicationManagerDatabase =
"asterisk";
        static private string ApplicationManagerUser = "
MySqlo User";
        static private string ApplicationManagerPassword = "
MySql Pass";

        /// <summary>
        /// LogmeNow.
        /// </summary>
        /// <param name="UserName">Current user.</param>string
        /// <param name="actions">What was
```

```
happening.</param>string
        /// <param name="Variables">The issue at
hand.</param>string
        /// <param name="PageId">What page is the user
on.</param>string
        /// <remarks>
        ///Where we can send logs to the database
        ///Because end-users lie
        /// </remarks>

        static public void LogmeNow(string UserName, string
actions, string Variables, string PageId)
        {
            string strSQL = "INSERT INTO Logging (UserName,
Action, Variables, PageID) " +
             "VALUES('" + UserName + "','" + actions + "','" +
Variables + "','" + PageId + "')";
            string strProvider = "Data Source=" +
LoggingManagerHost + ";Database=" + LoggingManagerDatabase +
";User ID=" + LoggingManagerUser + ";Password=" +
LoggingManagerPassword;
            MySqlConnection mysqlCon = null;
            try
            {
                mysqlCon = new MySqlConnection(strProvider);
                mysqlCon.Open();
                if (mysqlCon.State.ToString() == "Open")
                {
                    MySqlCommand mysqlCmd = new
MySqlCommand(strSQL, mysqlCon);
                    mysqlCmd.ExecuteNonQuery();
                }
            }
            catch (Exception er)
            {
                //This tells the program to log errors
                Database.ErrorLogmeNow("LogmeNow", er.Message,
UserName, "DataBase Class");
                mysqlCon.Close();
            }
            mysqlCon.Close();
        }
        /// <summary>
        /// ErrorLogmeNow.
        /// </summary>
        /// <param name="EFunction">Current Function that was
called.</param>string
        /// <param name="EError">What was
```

```
happening.</param>string
        /// <param name="EUser">The current
user.</param>string
        /// <param name="PageName">What page is the user
on.</param>string
        /// <remarks>
        /// We can log errors here
        /// Again End-Users lie
        /// </remarks>

        static public void ErrorLogmeNow(string EFunction,
string EError, string EUser, string PageName)
        {
            string strSQL = "INSERT INTO ErrorLog(Function,
Error, User, Page) VALUES('" + EFunction + "', '" + EError +
"', '" + EUser + "', '" + PageName + "'); ";
            string strProvider = "Data Source=" +
LoggingManagerHost + ";Database=" + LoggingManagerDatabase +
";User ID=" + LoggingManagerUser + ";Password=" +
LoggingManagerPassword;
            MySqlConnection mysqlCon = null;
            try
            {
                mysqlCon = new MySqlConnection(strProvider);
                mysqlCon.Open();
                if (mysqlCon.State.ToString() == "Open")
                {
                    MySqlCommand mysqlCmd = new
MySqlCommand(strSQL, mysqlCon);
                    mysqlCmd.ExecuteNonQuery();
                }
            }
            catch (Exception er)
            {
                mysqlCon.Close();
                Database.ErrorLogmeNow("ErrorLogmeNow",
er.Message, EUser,"DataBase Class");
            }
            mysqlCon.Close();
        }
        /// <summary>
        /// GetAllNums.
        /// </summary>
        /// <returns>The DataSet that we will bind
too.</returns>DataSet
        /// <remarks>
        /// Simple way to get all of the customer Id's
        /// </remarks>
```

```
        static public DataSet GetAllNums()
        {

            List<string> CurrentStatus = new List<string>();
            string strSQL = "CALL GetCustomerInfo()";
            string strProvider = "Data Source=" +
ApplicationManagerHost + ";Database=" +
ApplicationManagerDatabase + ";User ID=" +
ApplicationManagerUser + ";Password=" +
ApplicationManagerPassword;
            MySqlConnection mysqlCon = null;
            DataSet ds = new DataSet();
            try
            {
                mysqlCon = new MySqlConnection(strProvider);
                mysqlCon.Open();
                if (mysqlCon.State.ToString() == "Open")
                {
                    MySqlCommand mysqlCmd = new
MySqlCommand(strSQL, mysqlCon);
                    MySqlDataAdapter da = new
MySqlDataAdapter(mysqlCmd);

                    da.Fill(ds);
                }
            }
            catch (Exception er)
            {
               Database.ErrorLogmeNow("GetAllNums",
er.Message, "DataFunction", "DataBase Classes");
                mysqlCon.Close();
            }
            mysqlCon.Close();
            DataSet Rds = null;
            Rds = ds;
            return Rds;
        }
        /// <summary>
        /// GetCustomerInfo.
        /// </summary>
        /// <param name="CustomerId">The Customer ID,
generally the main phone number for the
customer.</param>string
        /// <returns>The DataSet that we will bind
too.</returns>DataSet
        /// <remarks>
        /// Where we get the information from the customer
```

```
        /// </remarks>

        static public DataSet GetCustomerInfo(string
CustomerId)
        {
            string strSQL = "CALL GetCustomerInfo();";
            string strProvider = "Data Source=" +
ApplicationManagerHost + ";Database=" +
ApplicationManagerDatabase + ";User ID=" +
ApplicationManagerUser + ";Password=" +
ApplicationManagerPassword;
            MySqlConnection mysqlCon = null;
            DataSet ds = new DataSet();
            try
            {
                mysqlCon = new MySqlConnection(strProvider);
                mysqlCon.Open();
                if (mysqlCon.State.ToString() == "Open")
                {
                    MySqlCommand mysqlCmd = new
MySqlCommand(strSQL, mysqlCon);
                    MySqlDataAdapter da = new
MySqlDataAdapter(mysqlCmd);
                    da.Fill(ds);
                }
            }
            catch (Exception er)
            {
                Database.ErrorLogmeNow("GetCustomerInfo",
er.Message, "DataFunction", "DataBase Classes");
                mysqlCon.Close();
            }
            mysqlCon.Close();
            DataSet Rds = null;
            Rds = ds;
            return Rds;

        }
        /// <summary>
        /// GetCustomerNumbers.
        /// </summary>
        /// <param name="CustomerId">The Customer ID,
generally the main phone number for the
customer.</param>string
        /// <returns>The DataSet that we will bind
too.</returns>DataSet
        /// <remarks>
        /// Gets all of the customer numbers associated with a
```

```
customer's ID
        /// </remarks>

        static public DataSet GetCustomerNumbers(string
CustomerId)
        {
            string strSQL = "CALL
GetCustomerNumberFromAccountCode('" + CustomerId + "');";
            string strProvider = "Data Source=" +
ApplicationManagerHost + ";Database=" +
ApplicationManagerDatabase + ";User ID=" +
ApplicationManagerUser + ";Password=" +
ApplicationManagerPassword;
            MySqlConnection mysqlCon = null;
            DataSet ds = new DataSet();
            try
            {
                mysqlCon = new MySqlConnection(strProvider);
                mysqlCon.Open();
                if (mysqlCon.State.ToString() == "Open")
                {
                    MySqlCommand mysqlCmd = new
MySqlCommand(strSQL, mysqlCon);
                    MySqlDataAdapter da = new
MySqlDataAdapter(mysqlCmd);
                    da.Fill(ds);
                }
            }
            catch (Exception er)
            {
                Database.ErrorLogmeNow("GetCustomerNumbers",
er.Message, "DataFunction", "DataBase Classes");
                mysqlCon.Close();
            }
            mysqlCon.Close();
            DataSet Rds = null;
            Rds = ds;
            return Rds;

        }
        /// <summary>
        /// GetAllCustIfo.
        /// </summary>
        /// <param name="CustomerId">The Customer ID,
generally the main phone number for the
customer.</param>string
        /// <returns>The DataSet that we will bind
too.</returns>DataSet
```

```
        /// <remarks>
        /// Gets all the inforation of a customers ID
        /// </remarks>

        static public DataSet GetAllCustIfo(string CustomerId)
        {
            string strSQL = "CALL GetAllCustInfo('" +
CustomerId + "');";
            string strProvider = "Data Source=" +
ApplicationManagerHost + ";Database=" +
ApplicationManagerDatabase + ";User ID=" +
ApplicationManagerUser + ";Password=" +
ApplicationManagerPassword;
            MySqlConnection mysqlCon = null;
            DataSet ds = new DataSet();
            try
            {
                mysqlCon = new MySqlConnection(strProvider);
                mysqlCon.Open();
                if (mysqlCon.State.ToString() == "Open")
                {
                    MySqlCommand mysqlCmd = new
MySqlCommand(strSQL, mysqlCon);
                    MySqlDataAdapter da = new
MySqlDataAdapter(mysqlCmd);
                    da.Fill(ds);
                }
            }
            catch (Exception er)
            {
                Database.ErrorLogmeNow("GetCustomerNumbers",
er.Message, "DataFunction", "DataBase Classes");
                mysqlCon.Close();
            }
            mysqlCon.Close();
            DataSet Rds = null;
            Rds = ds;
            return Rds;

        }
        /// <summary>
        /// GetCustomerVM.
        /// </summary>
        /// <param name="CustomerId">The Customer ID,
generally the main phone number for the
customer.</param>string
        /// <returns>The DataSet that we will bind
too.</returns>DataSet
```

```
        /// <remarks>
        /// Gets the voice mail for a customers phone number
        /// </remarks>

        static public DataSet GetCustomerVM(string CustomerId)
        {
            string strSQL = "CALL GetVMInfo('" + CustomerId +
"');";
            string strProvider = "Data Source=" +
ApplicationManagerHost + ";Database=" +
ApplicationManagerDatabase + ";User ID=" +
ApplicationManagerUser + ";Password=" +
ApplicationManagerPassword;
            MySqlConnection mysqlCon = null;
            DataSet ds = new DataSet();
            try
            {
                mysqlCon = new MySqlConnection(strProvider);
                mysqlCon.Open();
                if (mysqlCon.State.ToString() == "Open")
                {
                    MySqlCommand mysqlCmd = new
MySqlCommand(strSQL, mysqlCon);
                    MySqlDataAdapter da = new
MySqlDataAdapter(mysqlCmd);
                    da.Fill(ds);
                }
            }
            catch (Exception er)
            {
                Database.ErrorLogmeNow("GetCustomerNumbers",
er.Message, "DataFunction", "DataBase Classes");
                mysqlCon.Close();
            }
            mysqlCon.Close();
            DataSet Rds = null;
            Rds = ds;
            return Rds;

        }
        /// <summary>
        /// ChangeCustomerVMPass.
        /// </summary>
        /// <param name="CustomerMB">The Customer's New mail
box number.</param>string
        /// <param name="NewPass">The Customer' new
password.</param>string
        /// <remarks>
```

```
        /// Changes a customers voicemail password.
        /// </remarks>

        static public void ChangeCustomerVMPass(string
CustomerMB, string NewPass)
        {
            try
            {
            string strProvider = "Data Source=" +
ApplicationManagerHost + ";Database=" +
ApplicationManagerDatabase + ";User ID=" +
ApplicationManagerUser + ";Password=" +
ApplicationManagerPassword;
            MySqlCommand cmd = new
MySqlCommand("ChangeVMPAss", new
MySqlConnection(strProvider));
            cmd.CommandType = CommandType.StoredProcedure;
            cmd.Parameters.Add(new MySqlParameter("NewPW",
NewPass));
            cmd.Parameters.Add(new MySqlParameter("CustMB",
CustomerMB));
            cmd.Connection.Open();
            cmd.ExecuteNonQuery();
            cmd.Connection.Close();
            }
            catch (Exception er)
            {
                string Er = er.Message;
                Database.ErrorLogmeNow("ChangeCustomerVMPass",
er.Message, "DataFunction", "DataBase Classes");

            }
            return;

        }
        /// <summary>
        /// CreateCustomerVM.
        /// </summary>
        /// <param name="CustomerAccount">The Customer's ID
usually the main phone number.</param>string
        /// <param name="CustomerNum">The Customer's phone
number.</param>string
        /// <param name="MBNum">The Customer's mail box
number.</param>string
        /// <param name="MBPass">The Customer's mail box
password.</param>string
        /// <remarks>
        /// Createa new voicemail box for a customer
```

```
        /// </remarks>

        static public void CreateCustomerVM(string
CustomerAccount, string CustomerNum, string MBNum,string
MBPass)
        {
            try
            {
                string strProvider = "Data Source=" +
ApplicationManagerHost + ";Database=" +
ApplicationManagerDatabase + ";User ID=" +
ApplicationManagerUser + ";Password=" +
ApplicationManagerPassword;
                MySqlCommand cmd = new
MySqlCommand("CreateVMBox", new MySqlConnection(strProvider));
                cmd.CommandType = CommandType.StoredProcedure;
                cmd.Parameters.Add(new
MySqlParameter("IAccountCode", CustomerAccount));
                cmd.Parameters.Add(new MySqlParameter("IName",
CustomerNum));
                cmd.Parameters.Add(new MySqlParameter("IBox",
MBNum));
                cmd.Parameters.Add(new MySqlParameter("IPass",
MBPass));
                cmd.Connection.Open();
                cmd.ExecuteNonQuery();
                cmd.Connection.Close();
            }
            catch (Exception er)
            {
                string Er = er.Message;
                Database.ErrorLogmeNow("CreateCustomerVM",
er.Message, "DataFunction", "DataBase Classes");

            }
            return;

        }
        /// <summary>
        /// CreateCustomer.
        /// </summary>
        /// <param name="CustomerAccount">The Customer's ID
usually the main phone number.</param>string
        /// <param name="CustomerNum">The Customer's phone
number.</param>string
        /// <param name="Description">The Customer's
name.</param>string
        /// <remarks>
```

```
        /// Creates a new customer and or a new customer phone
number
        /// </remarks>

        static public void CreateCustomer(string
CustomerAccount, string CustomerNum, string Description)
        {
            try
            {
                string strProvider = "Data Source=" +
ApplicationManagerHost + ";Database=" +
ApplicationManagerDatabase + ";User ID=" +
ApplicationManagerUser + ";Password=" +
ApplicationManagerPassword;
                MySqlCommand cmd = new
MySqlCommand("CreateNewPhone", new
MySqlConnection(strProvider));
                cmd.CommandType = CommandType.StoredProcedure;
                cmd.Parameters.Add(new MySqlParameter("IName",
CustomerNum));
                cmd.Parameters.Add(new
MySqlParameter("IAccountCode", CustomerAccount));
                cmd.Parameters.Add(new
MySqlParameter("IDescription", Description));
                cmd.Connection.Open();
                cmd.ExecuteNonQuery();
                cmd.Connection.Close();
            }
            catch (Exception er)
            {
                string Er = er.Message;
                Database.ErrorLogmeNow("CreateCustomer",
er.Message, "DataFunction", "DataBase Classes");

            }
            return;

        }



    }
}
```

Bonus

Bonus for getting the current user status. I didn't add this code to the program but descided to place it here for you to play with on your own.

With this routine if you call it from C# you can then bind it to a gridview and display the current status of the extensions on the system.

MySqll routine.

```
CREATE DEFINER=`root`@`%` PROCEDURE `User_Status`()
LANGUAGE SQL
NOT DETERMINISTIC
CONTAINS SQL
SQL SECURITY DEFINER
COMMENT ''
BEGIN
SELECT Description AS "Customer-Name",
name AS Number,lastms,
useragent AS "Phone-Type"
FROM sip_buddies ORDER BY Description;
END
```

Add to Classes:

```
        static public DataSet Getnewstatus()
        {
            List<string> CurrentStatus = new List<string>();
            string strSQL = "CALL User_Status();";
            string strProvider = "Data Source=" +
ApplicationManagerHost + ";Database=" +
ApplicationManagerDatabase + ";User ID=" +
ApplicationManagerUser + ";Password=" +
ApplicationManagerPassword;
            MySqlConnection mysqlCon = null;
            DataSet ds = new DataSet();
            try
            {
                mysqlCon = new MySqlConnection(strProvider);
                mysqlCon.Open();
```

```
                if (mysqlCon.State.ToString() == "Open")
                {
                    MySqlCommand mysqlCmd = new
MySqlCommand(strSQL, mysqlCon);
                    MySqlDataAdapter da = new
MySqlDataAdapter(mysqlCmd);

                    da.Fill(ds);
                }
            }
            catch (Exception er)
            {
                // DataBase.ErrorLogmeNow("GetQueueWait",
er.Message, "DataFunction", "DataBase Classes");
                mysqlCon.Close();
            }
            mysqlCon.Close();
            DataSet Rds = null;
            Rds = ds;
            return Rds;

        }
```

Add to webpage (Note you have to add AjaxControlToolKit to references either by using Nuget or manually):

```
<asp:UpdatePanel ID="UpdatePanel1" runat="server">
            <ContentTemplate>
<br />
                <asp:GridView ID="CurrentStatus"
runat="server" OnRowDataBound="CurrentStatus_RowDataBound"
BackColor="White" BorderColor="#993399" BorderStyle="Solid"
horizontalalign="center" style="text-align: center"
Width="90%">

<AlternatingRowStyle BackColor="#00CCFF" />
                                                    <HeaderStyle
BackColor="#0066FF" />
                </asp:GridView>
                <br />
                <asp:Timer ID="Timer1" runat="server"
Interval="3000" OnTick="Timer1_Tick">
                </asp:Timer>
                <br />
            </ContentTemplate>
        </asp:UpdatePanel>
```

Add this to code behind:

```
        protected void Timer1_Tick(object sender, EventArgs e)
        {

            CurrentStatus.DataSource =
Database.Getnewstatus();
            CurrentStatus.DataBind();
        }
        //CurrentStatus_RowDataBound
        protected void CurrentStatus_RowDataBound(object
sender, GridViewRowEventArgs e)
        {
            if (e.Row.RowType == DataControlRowType.DataRow)
            {
                int x = 0;

                Int32.TryParse(e.Row.Cells[2].Text, out x);
                if (e.Row.Cells[2].Text == "-1")
                {
                    e.Row.Cells[2].ForeColor =
System.Drawing.Color.Red;
                }
                 else if (e.Row.Cells[2].Text == "0")
                 {
                     e.Row.Cells[2].ForeColor =
System.Drawing.Color.Blue;
                 }
                else if (x > 90)
                {

                    e.Row.Cells[2].ForeColor =
System.Drawing.Color.Violet;
                }
                else
                {
                    e.Row.Cells[2].ForeColor =
System.Drawing.Color.Green;
                }
            }

        }
```

ABOUT THE AUTHOR

I am Terry, I have 20 + years in technology including VoIP, Networking and Software design. I owned a technology company for 18 years working with large hospitals, government agencies, businesses and the general population.

I generally tell things like they are while using plain language. I feel that if the words have to be steam cleaned and pressed to be on pages then they aren't worth reading.

Enjoy the books and the different style of writing. There are more to come.

If you find yourself wanting consulting or further information, then feel free to email me direct at AsteriskHA@3states.net any time and I will do my best to answer your questions ASAP

www.ingramcontent.com/pod-product-compliance
Lightning Source LLC
Chambersburg PA
CBHW070222190526
45169CB00001B/49

* 9 7 8 1 5 3 0 1 6 9 3 4 4 *